Terrorism, Technology and Apocalyptic Futures

Maximiliano E. Korstanje

Terrorism, Technology and Apocalyptic Futures

Maximiliano E. Korstanje
University of Palermo
Buenos Aires, Argentina

ISBN 978-3-030-13384-9   ISBN 978-3-030-13385-6 (eBook)
https://doi.org/10.1007/978-3-030-13385-6

Library of Congress Control Number: 2019933127

© The Editor(s) (if applicable) and The Author(s), under exclusive licence to Springer Nature Switzerland AG 2019
This work is subject to copyright. All rights are solely and exclusively licensed by the Publisher, whether the whole or part of the material is concerned, specifically the rights of translation, reprinting, reuse of illustrations, recitation, broadcasting, reproduction on microfilms or in any other physical way, and transmission or information storage and retrieval, electronic adaptation, computer software, or by similar or dissimilar methodology now known or hereafter developed.
The use of general descriptive names, registered names, trademarks, service marks, etc. in this publication does not imply, even in the absence of a specific statement, that such names are exempt from the relevant protective laws and regulations and therefore free for general use.
The publisher, the authors and the editors are safe to assume that the advice and information in this book are believed to be true and accurate at the date of publication. Neither the publisher nor the authors or the editors give a warranty, express or implied, with respect to the material contained herein or for any errors or omissions that may have been made. The publisher remains neutral with regard to jurisdictional claims in published maps and institutional affiliations.

Cover image © Alex Linch / shutterstock.com

This Palgrave Macmillan imprint is published by the registered company Springer Nature Switzerland AG.
The registered company address is: Gewerbestrasse 11, 6330 Cham, Switzerland

This book is dedicated to the memory of Pablo Isaac Kaplan, and the Kaplan family in these difficult times.

Preface

Writing a book about the apocalypse theory is not easy, because the meanings and imaginings of apocalypse suffer different variations on time and culture. However, we are living at the borders of the turn of the millennium, just after the 2000s. Each 1000 years, societies experience anxieties and fears that are often articulated in a set of narratives known to anthropologists as "millenarianism". The imminent end of the world is given not only by human greed but also by excesses in the administration of the world. These cycles are accompanied by a sentiment of hope, a renovation which is aimed at enhancing social cohesion (Skoll & Korstanje, 2014; Wojcik, 1997). It is important not to lose the sight of the fact that in these discourses, technology plays a crucial role accelerating the moral decline that corrupted mankind. Although technology is originally designed to make the world a safer place, under some conditions, it accelerates its end (Greenberg, Rabkin, & Olander, 1983). Doubtless, the technological breakthrough over the recent years has introduced the man to a new (virtual) world. As Slavoj Žižek remarks, terrorism not only woke up the West from its slumber but also destroyed the wonderland of consumption, a type of virtual world characterized by the excess of technology and simulation (Žižek, 2015). In the Matrix Saga, Morpheus interrogates Neo to make a decision. Thomas Anderson is a computer programmer who lives a mediocre and routine life. Simultaneously, he lives a double life as the hacker Neo. He is approached by another hacker, Trinity, who introduces him to Morpheus. In a club room, Morpheus says to Neo that he needs to choose between two options. By taking the red or the blue pill, Neo will reach two different realities. While the blue pill will plant him in his daily boring life, the red will show the truth about the world and the Matrix. When Neo swallows the red pill, his reality dilutes. Once restored, Neo is educated about the new life of humans. In the twenty-first century, humans blocked the sun to deter the advance of machines. As a retaliation, machines enslaved humans, harvesting them as forms of energy. The Matrix is no other thing than a (controlled) simulation of the desired world while humans are placed in pods to provide energy to the machines. Matrix is a story where the creature rages against its creator enabling a world of prophecies and mythical conflagrations and interrogating the condition of human existence. Matrix reveals the mythical fear in adopting the technology, citing Weber, as an iron cage (Ross, 2017). It is safe to say the mythical plot of Matrix inspired countless studies and works which theorize about the ideological nature of technology and the impossibility of the real knowledge (Irwin, 2005), the roots of skepticism (Erion & Smith) or the dichotomy between living a real or an alienated life (Griswold, 2005) among many other themes. In his introductory chapter (in the book *The Matrix and Philosophy*), William Irwin calls the attention to the commonalities of Neo and Socrates. He writes:

> Neo is on a mission to save the human race from unwitting enslavement to artificial intelligence. Socrates too is on a mission, a mission from (the) God (Apollo), delivered via the Oracle of Delphi to his friend Chaerephon. His mission should he choose to accept it, is to wake up the people of his hometown, Athens. (Irwin, 2005: 6)

In the same way, the Oracle tells Morpheus a chosen hero (the One) will emancipate mankind from the junk of the Matrix. As the mythical narrative of the virgin maiden who accepts the God's wish, Neo accepts his destiny—though horrified of the world he faces. The life of the hero is fraught with risks, dangers, obstacles and even crime, but his outstanding character prevails. Unlike Neo, Cypher is exhausted of living in a grim world and sacrifices his liberty (making a pact with Agent Smith) to be reconnected to the Matrix.

> Cypher agrees to lead Smith to Morpheus in exchange for a new life as the wealthy, famous actor inside the Matrix. Cypher knows that the Matrix is not real, but he believes that he can make his life better by simply ignoring this and retreating back into a pleasant world of illusory fantasy. (p. 25)

Such a dichotomy, between truth and pleasure, is often encapsulated into the logic of the apocalypse landscape. God disposed of the Eden for the man who—by the action of greed or mistrust—betrayed him. Mankind is punished to live in a world of frustrations, need and deprivation, accepting labor as the only redeemable option. In this new reality, the labor and technology play a leading role helping men to control the environment, but to some extent, given his heart poisoned by rage and envy, God will purify the earth disposing of an expiatory disaster (like a flood or a great fire). As Andrew Feenberg acknowledged, technology is a double-edged sword. Through the years, social scientists developed a critical stance against technology, considered as the source of totalitarianism or the end of democracy. Today, technology poses a difficult dilemma (named as subversive rationalization) to man, which is expressed as follows:

> Rationalization in our society responds to a particular definition of technology as a means to the goals of profit and power. A broader understanding of technology suggests a very different notion of rationalization, based on responsibility for the human and natural contexts of technical action. I call this subversive rationalization because it requires technological advances that can be made only in opposition to the dominant hegemony. (Feenberg, p. 20)

Here, it is important to discuss two important ideas. Democracy and the respect for ethnic minorities are subordinated to the dogma of security when people feel fear. In the culture of fear or the days of terrorism, technology is amply used to limit the individual liberties of lay-citizens. At least, Snowden's case evidences the arbitrariness of the US government to violate the rights of privacy of Americans, as well as other human rights violations committed in Supermax prisons today. David Altheide (2014) refers to this event as "the triumph of fear", while Lyon (2015) called it "Snowden's oxymoron". Both ascribe to the notion that there is danger in democratic institutions ceding to the populist demands of security. Even though, terrorism represents a serious threat for many Western governments—if not a priority—this does not mean that the individual liberties may be suspended. In some respect, terrorism and the apocalypse theory are inextricably intertwined. Both depart from a fundamentalist viewpoint of the world where the evil should be defeated, and what is more important, both operate from a general belief (a prophecy, a closed narrative) where the Other's suffering should be tagged behind the collective well-being. In fact, terrorism and the millenarianism it generates have similar dynamics oriented to neglect the Otherness. This is exactly what happens in the days of Thana-capitalism and the morbid consumption, a new stage of capitalism where death is mediatized as the main commodity (Korstanje, 2016). These are the ideas that drew this book, some philosophical prerogative such as fear, terrorism, democracy, torture, the apocalypse theory and, of course, technology. The prophecy and disasters are inextricably intertwined. The elements of prophecies are hard to be defined, but surely ecstasy occupies a central position in the formation of prophetical narratives. In view of this, as Blenkinsopp highlights, the message, which often alerted on an imminent disaster or external attack, was given by the Gods invoking the needs of a previous verification (process of authentication). This happens because the problem lies in the nature of the prophecy as a disruptive text which contradicts or confronts the status quo, but to some extent:

> More important than the form is, of course, the content of the message delivered to the king. Most of them have to do with military affair-warning about revolt and the possibility of assassination, injunctions against undertaking certain expeditions or entering into certain alliances. In some instances the warning—e.g about fortifying a gate or not rebuilding a house—are accompanied by threats of unpleasant consequences. (Blenkinsopp, 1984: 57)

As a political artifact of domination, the prophecy interrogates the present from a near or distant future, subordinating the social order to the desires of the ruler, but what is more important, giving lessons that help community to face high levels of uncertainty. The first introductory chapter examines the anthropological theory about scatology and the end of the world. As myth-builders, we, the humans, move in a world of uncertainties. Far from what positivism precluded, myths are moral guidelines situated in an atemporal time, where Gods and humans coexisted in peace. As representational archetypes of the world and its limitations, myths say much about how practical problems should be addressed. The chapter debates the problem of evilness, as well as the obsession of modern culture for the bottom-days. As debated above, these narratives describe a declining situation where the introduction of technology corrupted the human's soul. Through the reading of authoritative voices such as Mircea Eliade, Lévi-Strauss, Mary Douglas and Joseph Campbell, we accompany readers through a fascinating world that helps them understand the contemporanean society from a new angle. Over years, not only was the fear politically manipulated to impose policies that otherwise would not be accepted, but the idea of "anti-Christ" was also adjusted to demean political adversaries. The apocalypse theory exhibits the philosophical quandaries between the human existence (which is open to the future) and the prophecy (closed to what cannot be changed). This would be exactly the dilemma of puritanism that has been discussed in the Weberian texts. The second chapter starts from the same premise, where, interrogating on the essence of zombie culture, we lay the foundations to construct a bridge between the precaritized reality of the workforce and the erratic zombies. Based on the novel *World War Z*, the chapter analyzes the passage of zombie stories extracted from the post-colonial Haiti to a global world where the Other's death gives pleasure to the survivors. Max Brooks' plot reveals three significant points—deciphered in the chapter. First, there is a clear rivalry between the official hero who embodies the power of law and order and the outlaw hero who confronts the evil to protect humanity. While the former does what should be done, but without any practical result, the latter breaks the law to struggle heroically to avoid humanity's destruction. Second, the plot retreats an apocalyptic virus surfaced in the uncivilized "China". Democracy, at some point, opens the doors to participation and liberty which is a vital step to deal with risks. Totalitarian governments—like China, Cuba or North Korea—are not only insensitive to what their citizens need but also fail systematically to locate and eradicate global risks. Third, the total war yielded by zombies entails the end of modern politics or the Hobbesian state as we know it. The evolution marks progress, which implies the eradication of other species. Like jihadists who loom in the European cities, though they—at some moment—looked like us, now they eat us! The success of zombies in invading the world reminds that our species is in jeopardy. Following this, the third chapter ignites a hot debate revolving around the advance of more radicalized terrorism and the death of hospitality in the West. Through the years, the theme of abortion has found a place on the agenda of many countries. Although the abortion law was already present in France, the US and the UK, in some other countries, the law was recently sanctioned just after the 2000s. This chapter deals with the problem of hospitality in the days of terrorism. From different angles, this chapter explores the question of abortion and the ideas of hospitality and multiculturalism, points that modern terrorism has instilled as necessary debates. We hold the thesis that the modern self has serious problems to understand the alterity when it confronts the own desires. The right of legal abortion and the surface of radicalized feminism should be framed as a decline of hospitality, a type of anti-hospitality imposed by modern terrorism and the culture of fear.

The 9/11 and the resulting War on Terror declared by Bush's administration not only changed geopolitics but also altered the life in the Western cities. The fourth chapter depicts the narrative circulating after 9/11 and the limitations of preemptive doctrine. Still further, the imposition of a fictional culture underpinned in a future that never takes place leads to accepting undemocratic practices on behalf of security. The arrival of conspiracy theory adjoined to the populist discourses that are oriented to eradicate the non-Western others (i.e., Muslims) is the result of 9/11 and the instrumentalization of terror and extortion as political means. Terrorism survives, after all, because the media provides the oxygen it needs (Eid, 2014).

In our book *The Rise of Thana Capitalism and Tourism*, we dangle how the risk society is changing to a new—more morbid—facet of capitalism where the Other's death is situated as a cultural criterion of entertainment. Viewed in TV programs, the Press, the macabre tours in new tourist destinations, video games and media entertainment, death and war are key commodities to be offered to thousands of users and viewers. Thana-capitalism exhibits this reality where the Other's suffering—far from being an opportunity for further empathy—represents the own happiness. Though originally we traced back the origin of Thana-capitalism to the Puritan faith and Anglo-Saxon England, the cradle of social Darwinism, a closer look suggests that the idea of a Messiah and the prophecy in Israel, as well as the figure of Holocaust where millions of innocents were killed by Nazis, occupies a central position. The archetype of holocaust emulated a sentiment of exemplarity which was for centuries present in Jewish culture. We not only explain how Zionism struggled against Judaism to impose in the philosophical debate about if Israel should become a nation-state but also describe how the victims are now the perpetrators.

The fifth chapter explores the genesis and evolution of Thana-capitalism, a new tendency recently emerged in the post industrial world. The classic tourist packages included sand and sun, while today the tourist taste mutates toward a morbid consumption. The sixth chapter explores the role played by the prophecy in the formation of the State of Israel. The term genocide was originally coined by Lemkin just after the horrendous crimes committed against innocent civilians in Nazi Germany. At that moment, the SS officials disposed of a systemic rationalized system of death which was oriented to domesticate and eradicate to the "inferior" or the undesired "Other". The concentration camps were space of torture, violence, death and mourning that marked the state of Israel forever. Today things have changed a lot, and the state of Israel is accused to violate the human rights in Palestine. While we review the discussion in senior lecturers as S Zizek, Richard Bernstein, Norman Finkelstein and Yakov Rabkin, we reconstruct the philosophical touchstone that led a nomad tribe to become as a state. This chapter deals not only with the sense of prophesy in Israel, but also it toys with the belief that the messianic idea of Messiah played a leading role in our appetite for consuming the Others' death.

The seventh chapter unveils the limitations of the global capitalism in understanding an ever-changing natural world. Though the climate change has come up as a major threat for mankind, no less true is the fact that the status quo recreates and circulates ideological narratives disposed to blame the victims as the main strategy. In this vein, poverty and poor communities are posed as the real challenges for the developed world.

From immemorial times, aborigines and ancient cultures were frightened of disasters because these events were seen as divine reprisals or punishment for their sins (as the Noah's Ark myth shows). Although the modern science introduced the instrumental thinking to understand disasters, improving the quality of life, it is no less true that capitalism obscured the diagnosis of scientists to protect the system. Blind to see the real problems of earth, today, capitalism offers distorted answers to the problems of climate change, migration and refugee crises, and even to the economic downturn. Based on the tactics of blaming "Others", the elite allude to poverty as the precondition toward humanitarian disasters.

We wish to thank Palgrave Macmillan (the content editors Anca Pusca and Kathleen Zingg) and the involving anonymous reviewers for the opportunity to publish this polemic work, as well as my family (my wife Maria Rosa Troncoso, my sons Ciro and Ben and my daughter Olivia, for the patience they displayed while this book was written). This book is dedicated to God who is the source of all hospitality, as well as migrants, refugees and asylum seekers who have to abandon their homes in quest of something better (the Eden). Most probably, like Adam and Eve, they found the doors of Paradise closed. Thanks to David Altheide, David Skoll, Adrian Scribano, Rodanthi Tzanelli and Freddy Timmerman for the talks and suggestions during this project.

Buenos Aires, Argentina   Maximiliano E. Korstanje

References

Altheide, D. L. (2014). The Triumph of Fear: Connecting the Dots About Whistleblowers and Surveillance. *International Journal of Cyber Warfare and Terrorism (IJCWT)*, 4(1), 1–7.

Blenkinsopp, J. (1984). *A History of Prophecy in Israel: From the Settlement in Land to the Hellenistic Period.* Philadelphia: The Westminster Press.

Eid, M. (Ed.). (2014). *Exchanging Terrorism Oxygen for Media Airwaves: The Age of Terroredia.* Hershey: IGI Global.

Erion, G., & Smith, B. (2005). Skepticism, Morality and the Matrix. In W. Irwin (Ed.), *The Matrix and Philosophy* (pp. 16–27). Chicago: Open Court.

Feenberg, A. (1995). Subversive Rationalization: Technology, Power and Democracy. In A. Feenberg & A. Hannay (Eds.), *The Politics of Knowledge* (pp. 3–22). Indianapolis: Indiana University Press.

Greenberg, M. H., Rabkin, E. S., & Olander, J. D. (Eds.). (1983). *The End of the World.* Carbondale: Southern Illinois University Press.

Griswold, C. (2005). Happiness and Cipher's Choice: Is Ignorance Bliss? In W. Irwin (Ed.), *The Matrix and Philosophy* (pp. 126–137). Chicago: Open Court.

Irwin, W. (2005). Computers, Caves, and Oracles: Neo and Socrates. In W. Irwin (Ed.), *The Matrix and Philosophy* (pp. 5–15). Chicago: Open Court.

Korstanje, M. E. (2016). *The Rise of Thana Capitalism and Tourism.* Abingdon: Routledge.

Lyon, D. (2015). *Surveillance After Snowden.* Cambridge: Polity Press.

Matrix. (1999). The Wanchowski Brothers. 136 Minutes. English, US, Warner Brothers.

Ross, C. (2017). *The Iron Cage: Historical Interpretation of Max Weber.* Abingdon: Routledge.

Skoll, G., & Korstanje, M. (2014). The Walking Dead and Bottom Days. *Antrocom: Online Journal of Anthropology*, 10(1), 11–23.

Wojcik, D. (1997). *The End of the World as We Know It: Faith, Fatalism, and Apocalypse in America.* New York: New York University Press.

Žižek, S. (2015). *The Universal Exception.* London: Bloomsbury.

Contents

1 Eschatology and the Theory of Apocalypse — 1

2 Interrogating on the Essence of the Zombie World — 19

3 The Undesired Other — 45

4 The War on Terror — 63

5 Tourism in the Days of Morbid Consumption — 81

6 Israel State, Genocide and Thana-Capitalism — 103

7 Disasters in the Society of Fear — 123

8 Conclusion — 143

Index — 151

CHAPTER 1

Eschatology and the Theory of Apocalypse

Introduction

The turn of the century has instilled many fears and anxieties in the public opinion. The proliferation of risks, associated to the climate crisis, disposed a climate of insecurity in the core of Western cities as never before. The term *risk society* coined by Ulrich Beck made much sense just after the 2000s. Beck envisaged a society where the logic of instrumentalism creates uncontemplated and apocalyptic risks which would very well place the society into jeopardy (Beck, 1992). The excess of instrumentality, which characterizes the late modernity, places all citizens in egalitarian conditions before the risk, as Beck infers. Not surprisingly, the attacks on the World Trade Center, which were planned by Saudi fundamentalist Osama Bin Laden and his radicalized cell Al-Qaeda, shocked the US and its allies, who formed an antiterrorism force. George W. Bush, who was ideologically educated and intellectually formed in the Pentecostalism, declared the "War on Terror" as an answer of a dormant Giant who had the responsibility to take care of the world against what he dubbed "the axis of evil" (Jacobson, 2008; Weisberg, 2008). The discourses and narratives circulating in post-9/11 contexts were impregnated of a much deeper eschatological logic, which were historically enrooted in the American tradition. Some voices even alerted on the connection of Nostradamus' prophecy, the rise of the third Anti-Christ and the 9/11. At a first glimpse, Bush's discourse was accompanied by "rad-cons" (radical conservatives) who strongly believed in the needs of the US to take direct intervention in the world (Walliss & Aston, 2011). As a dangerous

© The Author(s) 2019
M. E. Korstanje, *Terrorism, Technology and Apocalyptic Futures*,
https://doi.org/10.1007/978-3-030-13385-6_1

place, which opposes the ideals of democracy, rad-cons widely adhere to the theory that the non-Western world should be ethically controlled and domesticated, introducing democracy as the touchstone. This ideologically legitimated the two US-led invasions in Middle East, as well as the obsession struggling against terrorism as an efficient step in protecting the West (Berkowitz, 2004). In order for readers to gain further understanding of this complex setting, this introductory chapter deals with the nature of bottom-days theory and its connection in modern politics. The first section examines the importance of mythology and exegesis in the study of social world and the systematic reaction of positivism that considered them as "pseudo-science". Paradoxically, while a portion of sociology rejected mythology as scientific form of knowledge, ethnology and anthropology cemented their authorities in the fields. Far from being speculation or primitive discourses, ethnologists acknowledged that myths are moral guidelines which help society to overcome practical problems as the ancestors did. In other words, mythologies allow the cultural and social reproduction of society in the threshold of time. In second section, we put Mircea Eliade and his interesting intellectual legacy into the foreground. As he noted, myths are cultural constructions which are cyclically replicated in accordance to the cycles of production. The myths of apocalypse remind not only of the temporal nature of existence but also of the dichotomies of economies which alternate states of prosperities with crises. In the third section, we scrutinize in depth the seminal book, *The Fourth Turning: An American Prophecy*, which is originally authored by Strauss and Howe. As authors put it, one of the limitations of religion rests on the problem of "evilness". While religion accepts the existence of evil as a feasible possibility, the theory of apocalypse accelerates the times placing the final fight (against this looming evil) as a final solution or the last obstacle toward the prosperity of mankind. This seems to be exactly what captivates the attention of the masses and the reasons behind the adoption of "bottom-days" in the cinema industry. Thus, in the final section, we review the importance of these apocalyptic theories and discourses as platforms of cultural entertainment that today dominate the media.

Mythology and Social Sciences

Somehow, historically associated to certain stereotypes and prejudices enrooted in the Enlightenment, mythology and exegesis, which are the necessary methodologies to interpret myths, are not considered serious instruments in the construction of scientific knowledge (Hanen, Osler, & Weyant, 2006). In fact, some epistemologists as Karl Popper discouraged mythology as a serious scientific option based on his idea of falsifiability, which suggests that any theory seems to be useful only if it can be empirically tested, confirmed or neglected. In view of this, mythology, per his viewpoint, falls in the fields of *pseudo-sciences* because of the impossibility of being empirically tested (Popper, 1957, 2013). Needless to say, Popper overtly and systematically rejects the figure of induction centered on the fact that it leads very well to mere speculations.

As this backdrop, this position was the dominant paradigm during the days Joseph Campbell studied in Columbia University, where his Ph.D. dissertation in medieval literature was overtly neglected. Campbell was an authoritative voice in the study of myths and the interpretation of cultures despite the ungrounded criticism he received in life. Quite aside from this controversy, others discipline as ethnology and anthropology made of mythology their tugs of war, as well as a more than efficient instrument toward a better understanding of culture (Stocking, 1974). To put this slightly in other terms, while philosophy and sociology avoided discussing the strengths and weaknesses of mythology as a valid method of study, anthropology went precisely in the opposite direction. As Lévi-Strauss (1969) puts it, myths and the process of myth-building cannot be understood in isolation but integrated to a whole system. This was particularly innovating by the 1960s when he published *The Raw and the Cooked: Introduction of a Science of Mythology*. Lévi-Strauss laments the social sciences overlook the possibility to start a serious discussion revolving around myths. From the outset, his efforts were oriented to introduce readers toward a new science of mythology that explains the formation of cultures no matter what their economic matrixes. This means that the same logic that prevailed in the aboriginal mind can be found in the Westerners (Lévi-Strauss, 1966). In the anthropological fields, Lévi-Strauss was known and ultimately baptized as the father of structuralism. Regardless of his genius, he received the critique of another brilliant colleague, Mary Douglas. She convincingly argues that Lévi-Strauss' observations sound interesting but implausible. Her criticism is based on the fact that myths cannot be theoretically interpreted lest by the local practices observed in the field. Two structural myths may share commonalities, but this does not credit scientific correlation (Douglas, 1967).

> Levi Strauss unguardedly says that the units of mythological structure are sentences. If he took this statement seriously it would be an absurd limitation on analysis. But in fact, quite rightly, he abandons it at once, making great play with the structure underlying the meaning of a set of names. (Douglas, 1967: 50)

The multiple meanings and allegories each myth has, as Lévi-Strauss mistakenly recognizes, place structuralism in a conceptual gridlock. What is equally important, the correct understanding of myth cannot be dissociated by the practices that draw such interpretations. In view of this, Douglas not only wounds structuralism in its core, but also encapsulates myths into the possibility to change in the time (Douglas, 1967).

FitzRoy Somerset 4th, popularly known as Lord (Baron) Raglan, was a British soldier and amateur anthropologist who was originally motivated by the role of heroism in the contemporary culture. His erudition and passion for ancient history ushered him into the lives of different ancient heroes as Oedipus, Theseus, Heracles, Jason, Perseus and so forth. In his seminal book *The Hero: A Study of Tradition, Myth and Drama*, Raglan realizes that though unverified by the empirical information, myths—as the founding parents of anthropology—signal to abstract ideas which surely help society to understand the world. Myths often narrate complex ideas related to the creation, the life, the death and the bottom-days. Thinkers of the caliber of Durkheim, Mauss, Hubert or Malinowski—Raglan writes—agreed that myths exist not only in a savage community through the circulation of rituals and stories, but also in a modern community as historical facts that connect with a much deeper archetype of the hero. Such an archetype is not limited to aboriginals but remains enrooted in the Western civilization. This means that there is nothing different in the life of Siegfried as compared to Christ. Raglan patiently draws a wide conceptual framework to understand heroism as an intercultural reality. In his model, heroes follow the following traits:

1. The hero's mother comes from a royal kinship.
2. His father may be a king or a god.
3. Though reputed as the son of a god, he or she is forced to abandon home, most probably growing abroad until his or her adulthood.
4. We know little about his or her childhood.
5. He becomes a semi god or a king after defeating dragons, beasts or even confronting Gods.

6. He descends to the hell and resurrects, reminding that the human character can defy death.
7. He or she confronts the Gods to become a protector of mankind.

All the points enumerated above conform to the narratives of heroes in different stories, myths and legends. Raglan concludes that beyond the unhistorical nature of myths, they connect to a natural shared interest: the needs of transcendence.

> The thesis of this book is that the traditional narrative, in all its forms, is based not upon historical facts on the one hand or imaginative fictions on the other but upon dramatic ritual or ritual drama. I began by attempting to show that the belief that people have a natural interest in historical facts and a natural ability to transmit them is devoid of foundations. I then took a number of quasi-historical traditions and showed that there is no valid evidence for their historicity. I next gave the evidence for connecting the myth and the folk-tale with ritual, and for believing that hero-tale is derived from ritual and not from fact. (Raglan, 1956: 278)

As the argument given previously, Otto Rank (2013) traces back the nature of social behavior (which includes practices and beliefs) into what he dubbed as "elementary thoughts". Centered on the psychoanalytical theory, but profoundly interested by mythology, *he* found the possibility to create an archetype of heroism (following the efforts of Lord Raglan). Rank says that myths should be defined as collective construes which condition and determine the social mood. The sagas of heroes correspond with the needs of replacing the real father by a mediator mask, which interpellates the subject according to an imaginary world.

> The entire endeavor to replace the real father by a more distinguished one is merely the expression of the child's longing for the vanished happy time, when his father still appeared to be the strongest and greatest man, and the mother seemed the dearest and the most beautiful woman. The child turns away from the father, as he now knows him, to the father in whom he believed in his earlier years, his imagination being in truth only the expression of regret for this happy time having passed away. (Rank, 2013: 115)

This raises a more than an interesting question: are history and myth two sides of the same coin—so to speak—as Mircea Eliade thought?

Mircea Eliade and the Myth of Eternal Return

Over years, cultural theorists and philosophers have debated to what extent history and mythology should be separated in the academic circles. While history denotes the scientific inquiry of historical events as happened in the past, mythology alludes to an atemporal time which cannot be empirically demonstrated (Rickles, 2014). Mythology exerts a great influence in the aboriginal mind, as it was discussed in the earlier section, whereas history—which is often subject to carry out archival research—enthrones a so-called objectivity reflecting facts and events as they really occurred. Of course, for some scholars as Mircea Eliade, a Romanian historian who does not need the previous presentation, this conceptual dissociation seems to be a simplification. At a first glimpse, one of his most notable contributions was the differentiation between the profane and the sacred. Eliade said that any religion is manifested by the presence of *hierophany*, which means the emergence of the sacred. These hierophanies not only sort but also give meaning to the external world (Eliade, 1959). As his main entry in this discussion, Eliade understands that myths exhibit a total understanding with respect to a primordial past. In his book *Myth and Reality*, he holds the thesis that myths are more complex than fictional stories as some social scientists preclude. Eliade's studies start with some societies where myths are at the least living. In this respect, we must understand the structure and function of the myth in society, even in our contemporary society. For Eliade, humans are myth-makers no matter the culture or time. Ancient Greeks believed in Aquila, while Modern believers speak of a man-God they know as "Christ".

> Myth narrates a sacred history; it relates an event that took place in primordial Time, the fabled time of the beginning. In other words, myth tells how through the deeds of Supernatural Beings, a reality come to existence, be it the whole of reality, the Cosmos, or only a fragment of reality—an island, a species of plant, a particular kind of human behaviour, an institution. (Eliade, 1998: 5–6)

In Eliade's account, myth offers valid explanations and vivid guidelines on how the problems and threats have been faced by the founding parents in the pastime. In the same way, we—the modern—should deal with similar external risks. Myths contain all the necessary knowledge for the society to survive. As he puts it:

> For knowing the origin of an object, an animal, a plant, and so on is equivalent to acquiring a magical power over them by which they can be controlled, multiplied, or reproduced at will. (p. 15)

In his epistemological project, unlike Lévi-Strauss, the figure of myth contains the rationality of the society, which is projected by the desire to control the surrounding environment. This invariably leads to assumption that myths are not limited to aboriginal life, as they can be found in the modern culture as well. Myths share common elements that form the individual experience such as:

1. Myths certainly draw the ancient history of Supernatural Beings.
2. The stories narrated in myths are considered as true and sacred.
3. The myth is often related to the origins of the world.
4. Scrutinizing through the myth is a valid way of knowing the origin of things and the creation. In so doing, humans attempt to control the known world.

As the previous backdrop, when society is placed in jeopardy or is facing a climate of uncertainty, the collective rituals are performed to invoke what Eliade names as "the prestige of beginnings". Those who control the narratives revolving around the essential moment when the world was created are in a best position to impose their mandate to others who are far from changing the content of these stories.

> The myth briefly summarizes the essential moment of the Creation of the World and then goes on to relate the genealogy of the royal family or the history of the tribe or the history of the origin of sickness and remedies, and so on. In all these cases the origin myths continue and complete the cosmogonic myth. (ibid.: 36)

Following the question whether the myths of origin or the cosmogonic landscapes are of paramount importance in the construction of politics and culture, what should be debated is the role of millennialists or scatological myths.

This was particularly a topic Eliade addresses in his seminal work, *The Myth of Eternal Return: Cosmos and History*. He is moved to continue the old discussions that structuralism have left open. One of the points that distinguish the modern from the archaic man seems to be associated with how the cosmos is constructed. While the former is strongly attached to history, the latter is connected to the cosmic rhythms, as Eliade comments. The history of how the World was created should be preserved, as Eliade observes, and repeated to the next generations cyclically. As he discussed through his vast bibliography, myths should be defined as living guidelines that preserve exemplary models for all the activities men engage in. These exemplary models function as paradigms, which emphasizes the norms and habits that people should adopt and respect. Eliade believes that "archaic ontology" located in different cultures that makes certain events look real for premodern societies. These signs are related to the economic cycles of production in the original agrarian communities.

> The World that surrounds us, then, the world in which the presence and the work of man are felt—the mountains that he climbs, populated and cultivated regions, navigable rivers, cities, sanctuaries—all these have an extraterrestrial archetype, be it conceived as a plan, as a form, or purely and simply as a double existing on a higher cosmic level. (Eliade, 1959: 9)

The discoverers, conquerors or travelers took possession of unknown distant lands. They are appropriated through the sign of the language. When the settlers arrive to a new region, as Eliade brilliantly explains, the uncultivated country is symbolically indexed as an act of creation. In that way, the repetition of the primordial act—as the Gods did—is ensured. This paves the ways for the rise of a new dichotomy between the exemplary center, which emulates the temple of Gods and the periphery (where the anti-Gods live). The cycles of life are conveniently framed in the "annual expulsions" of demons which symbolize the disease and pain, and those rituals that articulate annual celebrations as the example of New Year show. His main thesis is that the time is regenerated through the articulation of cycles of scarcity and prosperity, which accompany mankind from its inception. These cycles emulate the sowing and cultivation of the lands, in the same way the myths of creation and destruction do. The cultivation of the unknown territory signals to those stories the mythical narratives indicate was created by the God (cosmogonic myths), while destruction (which is part of scatology) emulates the mowing of the land. Not surprisingly, the water and fire, which play a leading role in the mowing process, appear repeatedly through the countless scatological myths. The end of the world represents a period of darkness, where the world is not totally destroyed but renovated in a new cycle. Lastly, Eliade clarifies that the eternal return regulates not only the cycle of life and death, but also the means of productions molding the society in itself.

The Fourth Turning

It is safe to say that one of the most salient books in the fields of apocalypse and the sociology of bottom-days leads us to *The Fourth Turning* originally written by William Strauss and Neil Howe. In this editorial project, they hold the polemic thesis that originally pursued structuralism in the fields of anthropology, which means that history determines cyclically the human behavior. Events as they happen in our external world influence not only our behavior but also our habits. Understanding myths is a good option to grasp why the same events that alter the life of society finally recur.

> The reward of the historians is to locate patterns that recur over time to discover the natural rhythms of social experience. (Strauss & Howe, 1997: 2)

Put things in this way, Strauss and Howe identify three subtypes of time: (a) chaotic time, (b) cyclical time and (c) linear time. The first subtype is proper of hunters-and-gathers and other tribes that consider the time as chaotic, accepting that the history has no path. Events are randomly succeeded in reminding always that the world is inexpugnable for the human mind. Rather, the cyclical time signals to the rise of sedentary tribes, when the cycles of lives and planetary events such as solar years or diurnal rotations were celebrated in festivals and rituals. Although in this stage, the man tries to control the time, it takes a cyclical dynamic, which repeats itself in cohorts. Finally, the linear time corresponds with the modern intent to impose *lineralism* to administer the environment. This was exactly what happened with Europe once the Middle Age set the pace to modernity. The appearance of history, in the core of Western rationality, paved the ways for a new "urgency", which shaped the modern science. This represented a serious methodological limitation for social scientists who habitually overlook that history tends to repeat during a certain lapse of time. Rather, Europeans and Americans see themselves as unique, outstanding and finely located at the top of evolution with respect to the aboriginal organizations. As authors go on to say:

> They failed to realize that all generations were poised to enter new phases of life- and what, as they did, people up and down the life cycle would think and behave differently. (p. 18)

As the previous argument given, each life-cycle trajectory results in generations as well as emerging *archetypes* which are expressions of what Strauss and Howe dubbed as "enduring temperaments". There are four turnings (changes) that mark the moods, fear, experiences and cosmologies of society given the time. To a closer look, each turning engenders its own characteristic generation. The first turning (high) appeals to the *prophet generation*. A clear example of this dates back to the end of WWII and the baby-boomer generation, where the US witnessed a renaissance of the community life. Not only did people believe in hard work but the fear totalitarianism installed in the hearts of Americans also led them to embrace progress and growth as mainstream cultural values. The second turning known as *Awakening* refers to the *Nomad generation*. This era is distinguished by a passionate spiritualism that reacts when the civil order and the law are in jeopardy. The third generation, which is encapsulated in the third turning, comes when citizens and society suffer what C. Lasch called "the culture of narcissism". The individuality adjoined to the struggle for pleasure maximization gradually places the society in serious problems to identify the looming threats. The *hero generation* is born during the third turning which is baptized by Strauss and Howe as *the unraveling*. Lastly, the *crisis* (starting with the *artist generation*) can be understood as a decisive epoch where the civic order and the social institutions are in decline. Worried by a secular upheaval, the legitimacy of authorities is undermined to the extent a new civil order emerges. Fears, angst and the idea of the apocalypse or the end of the world—as we know—are stronger during *the artist generation*. Of course, since Westerners are educated to think the society in a unilineal time, which suggests that there is nothing like a return of social maladies, the idea of an end is more tragic and worrisome. In view of this, authors acknowledge we are coming through the crisis facet (or four turning) which is accelerated just after 2005.

The Theory of Apocalypse Explained with Clarity

In earlier sections, we explored with some detail the nature and function of cosmogonic myths, as well as the role played by scatology in the formation of culture. Here, rather, we delve into what specialists dubbed as "the theory of apocalypse", or in terms of Gary Wolfe (1983), the post-holocaust fiction. To some extent, the idea of apocalypse not only has changed in the threshold of time, but also has been commoditized by the consumerist society toward a new global product (Korstanje, 2016).

The fact is that while each culture develops the idea of an exemplary center which is prohibited to humans because of pride or the moral decline, it is equally true that from that moment on, civilizations try to recover their lost-paradise or grace through different acts of creation. The scatological myths speak of a world where the primitive state of nature re-emerges in order for humans to conciliate with Gods (Cohn, 1996). The idea of apocalypse reminds of not only the vulnerability of mankind, but also how greed, ambition and pride led a prosperous civilization to an inevitable end. The bottom-days articulate narratives of fear, which allude to an imminent apocalyptic scenario where the world as we know is completely placed upside down, with hopes which made the leap toward a new renovated era (Kumar, 1996; McGinn, 1996). As K. Kumar (1996) puts it, unlike others millenarianisms, the turn of the century installs an extreme sentiment of panic simply because the process of secularization, which postulated the end of after-life, impedes the adoption of hope as an alternative sentiment toward the idea the World would come to an end. Eric Rabkin (1986) contends that one of the aspects that characterizes the theory of apocalypse seems to be that the world is never fully destroyed. Rather, it is renovated through what he dubbed as "dark technology". To put this in bluntly, the world as we know it is not totally obliterated but certainly purged through the orchestration of divine plan. This evinces that humans, for many reasons, offended the Gods, and the purge restores the harmony in the universe. Once created, men were dotted of free-will, which suggests the possibility to make their own decisions. However, they gradually were seduced or tempted to defy Gods. A great variety of mythical narratives describe a world dominated by the man and technology, which contrasts to the natural order. To correct things, God disposes of the earth to be redeemed, sublimated into a new sanitized facet of spiritual existence. In view of this, the figure of exile exhibits the human impossibility to return to the lost paradise, but what is more important, it reminds the uncontrolled violence and human greed lead to the world destruction (Rabkin, 1986). Once the apocalypse takes place, humans are not eradicated but allegedly forced to live in a new condition.

As the previous backdrop, Gary Wolfe coins the term, *zero moment*, to theorize about a primitive state where humans struggle for basic survival. In this vein, Wolfe argues that the apocalypse prompts a radical reconfiguration of political order, which associates to the end of the state of civility and the constitutional law. The individual liberties as they are granted by the democratic society are not applied any longer, whereas a new aristocracy in a chaotic world surfaces. Whether technology prompts the rise of a civilizatory process, protecting the interests of the status quo, it accelerates the inevitable decline of the human race. The idea of *remaking zero* alludes to the needs of men to grasp their destiny, further increasing their understanding on their being in this new world. Technology becomes the touchstone of society, but at the same time its Aquila's heel. Wolfe goes on to say that:

> The promise inherent in the idea of remaking zero is certainly one of the reasons this genre has survived as long as it has, and in so many guises. On the simple level of narrative action, the prospect of a depopulated world in which humanity is reduced to a more elemental struggle with nature provides a convenient arena for the sort of heroic action that is constrained in the corporate, technological world that we know. (Wolfe, 1983: 4)

Ultimately, he enumerates some specific characters that are shared by the apocalyptic narratives in different cultures. The mythical hero starts a risky journey *through the wasteland* to liberate community or founded a new order. Regardless of the context, in the mythical narratives, heroes should face countless threats, but they are never brought to their knees. A whole portion of mythical landscapes centers on the fact that heroes live a lonely life when the disaster takes hit. Such isolation comes as a punishment in view of a primordial sin, crime or fault the hero involuntarily committed. Many of them even are marked or doomed by a much deeper sentiment of loneliness which leads them very well toward depression. Traveling so far to understand the reasons of the disaster, as well as their position there, heroes struggle not only to defeat the forces of evil but also to protect mankind in a new refounded order.

In his book, *Arguing the Apocalypse*, Stephen O'Leary (1998) says that the arrival of a new millennium always fascinated humanity. The approach of 2000 and this long-dormant fascination for the Armageddon captivated the attention of scholars as well as the public opinion with the same intensity.

> This scatological understanding can be seen in the age-old lament that describes the decline of morals in society: young people no longer respect their elders, while war and all kinds of immorality increase in direct correlation with the growth of humanity's knowledge and technical skills. (O'Leary, 1998: 5)

The above citation reveals two important aspects of the apocalypse. On the one hand, the humanity morally declines at the time the economic prosperity is paradoxically enthroned. On the other hand, technology plays an ambiguous role in expanding the civilizatory process but accelerating the prophetic vision of an ultimate destiny. This means that the end, which was predicted by the prophets and their visions, represents the dramatic (mythical) fight between god and evil. Having said this, as O'Leary notes, once the credibility of the ruling elite slumps down, the manipulation of prophecies seems to be a valid alternative solution. Since the nature of evil is situated as external to the community, any attempt to neutralize the devil is oriented to vulnerate the alterity.

> The discourses of conspiracy and apocalypse, therefore, are linked by a common function: each develops symbolic resources that enable societies to define and address the problem of evil. While conspiracy strives to provide spatial self-definition of the true community as set apart from the devils that surrounds us, apocalypse locates the problem of evil in time and looks forward to its imminent solution. (ibid.: 6)

With this in mind, O'Leary holds the thesis that apocalypse allows the construction of a common-shared sentiment of community whose function aims to social cohesion. This begs some questions the present chapter addresses: is apocalypse a necessary narrative to prevent the decline of social ties? What are the social forces that lead toward the fragmentation?

The Theory of Apocalypse and Cinema

The questions formulated in the introductory chapter guide partially the common-thread argument of this book. In a seminal work, entitled *Transmedia Storytelling and the Apocalypse*, Stephen Joyce (2018) argues that the end of civilization, at the least as it was created, captivates the attention of the masses as a real cultural entertainment. As a new genre, the post-apocalyptic consumption speaks of how viewers feel the end is nearer though in the comfort of the home. This ranges from the rise of hungry zombies to natural disasters and nuclear accidents. Joyce's thesis is that some deeper changes in the popular opinion have taken place, transforming the post-apocalyptic discourse as a cultural product which is widely consumed in the capitalist society. This cultural transformation creates a new demand, where the idea that the worst is coming feeds the needs of consuming the

"Other's suffering". Methodologically speaking, one of the limitations of post-apocalyptic worlds consists in the dispersion of the landscapes and the symbolic material they present. It is very hard to grasp an all-encompassing model that explains all apocalyptic discourses, as well as the individual appropriations behind them, as Joyce claims. What is clear, the plots bespeak of small groups or individual persons who struggle for surviving in extreme conditions, which may be caused or not by the action of man. The figure of technology occupies a central position in the post-apocalyptic scenarios. Sometimes technology serves as savior and sometimes as double-edge sword paving the ways for the rise of an unnamed threat that places mankind in the bias of annihilation. As Joyce notes:

> The post-apocalyptic is a portal fantasy in which the apocalyptic event functions as portal to an alternate world defined by the central narrative tensions of independence-dependence, progress-regression, utopia-dystopia, and the dominant motif of ambivalence about technology. (Joyce, 2018: 10)

As highlighted above, Joyce dangles that some sociological answers to the problem of apocalypse center on the fact that peoples sublimate their frustrations with the hope everything which causes pains will be exterminated. The question whether evangelicals, within the US, echo these types of demands appears to be one of the conceptual gridlocks scholars fall at the time of studying the issue. For some reason, liberal scholars, as Joyce adheres, erroneously think that religious leaders are fundamentalist terrorists which fuel their blood and hear with anger and hate. Joyce acknowledges that apocalypticism includes not only a leader who should be followed, but an enemy and final conflagration which stimulates the "us and them" logic. In a nutshell, Joyce contends what one of the features that identifies the post-apocalyptic world consists in the lack of hopes the good triumphs over the evil. The importance and acceptance of these narratives by the side of spectorship appear not to be given by the total destructions behind but by the mere possibility to bring the modern-man into another "imaginary world", Joyce concludes.

> The genre's growing complexity and evolving conventions, in turn, have attracted creative talents to explore its possibilities in prestige niches, thereby enhancing its reputation as an aesthetic form. The post-apocalyptic's emergence as an ideal form of transmedia storytelling is not a fortuitous accident but a process of genre-medium coevolution, in which new

ways of telling stories shape emergent genres, while the possibilities opened up by those genres also drive creators to find new ways of telling stories. (Joyce, 2018: 210)

After further review, Joyce's text saturates the discussion with rich empirical study-cases, analyzing not only post-apocalyptic plots but also the intersection between the bottom-days theory and modern politics. Though the book lacks a much deeper discussion around the nature of apocalypse, or at the least a review of the literature, it lays the foundations toward a new paradigm that explains why modern audiences are captivated and fascinated by this genre. Particularly interesting seems to be the fact that audiences are moving to "a morbid consumption" where the Other's suffering is systematically commoditized, packaged and gazed. This holds true for what we dubbed as "Thana-capitalism" (Korstanje, 2016), a moot point which will be touched in the next chapters. The present introductory section is a fresh insight that stresses the importance of myths in the formation of society and culture, as well as "the power of apocalypse" to guide and shape socially the individual behavior.

Conclusion

As discussed in this occasion through the articulation of different authoritative voices, myths are essentially of paramount importance for the reproduction of culture. As myth-builders, humans allude to archaic states of the consciousness where they lived in peace with the Gods. The introduction of sin, the work or the technology accelerated not only the moral decline of mankind but also their exile from Eden. From that moment onward, the figure of technology played an ambiguous role as a double-edge sword, which may spiritually boost or destroy humanity. The present chapter discusses to what extent the theory of apocalypse nourishes the millennial fears of the end of the world transforming them into a cultural phenomenon, which can be crystalized through the media consumption. This editorial project is a preliminary insight that reflects the intersection of mythology and politics keeping on the original legacies of Mircea Eliade, Lévi-Strauss, Mary Douglas and Joseph Campbell. In this field, anthropology and ethnology have much to say on the power of apocalypse and scatology to regulate modern politics. Over centuries, not only fear was politically manipulated to impose policies otherwise would be widely rejected but also the figure of "the anti-Christ" symbolizes the dichotomies between the human existence and the prophecy. Last but not least, the main argumentation—materialized through the different chapters that form this research—toys with the belief that behind the figure of the prophecy lies the needs of domesticating an uncertain future which is politically adjusted for the ruling elite to prevent the social change.

References

Beck, U. (1992). *Risk Society: Towards a New Modernity* (Vol. 17). London: Sage.
Berkowitz, P. (2004). Politicizing Reason. *Policy Review*, 1(127), 89.
Cohn, N. (1996). Upon Whom the Ends of the Ages Have Come. In M. Bull (Ed.), *Apocalypse Theory and the End of the World* (pp. 33–49). Oxford: Blackwell.
Douglas, M. (1967). The Meaning of Myth. With Special Reference to 'La Geste d'Asdiwal. In E. Leach (Ed.), *The Structural Study of Myth and Totemism* (pp. 49–69). London: Routledge.
Eliade, M. (1959). *The Sacred and the Profane: The Nature of Religion* (Vol. 144). New York: Houghton Mifflin Harcourt.
Eliade, M. (1998). *Myth and Reality*. Long Grove: Waveland Press.
Hanen, M., Osler, M., & Weyant, R. (2006). *Science, Pseudo-Science and Society*. Waterloo: Wilfrid Laurier University Press.
Jacobson, G. C. (2008). *A Divider, Not a Uniter: George W. Bush and the American People: The 2006 Election and Beyond*. New York: Longman Publishing Group.
Joyce, S. (2018). *Transmedia Storytelling and the Apocalypse*. Cham: Palgrave Macmillan.
Korstanje, M. E. (2016). *The Rise of Thana Capitalism and Tourism*. Abingdon: Routledge.
Kumar, K. (1996). Apocalypse, Millennium and Utopia Today. In M. Bull (Ed.), *Apocalypse Theory and the End of the World* (pp. 233–260). Oxford: Blackwell.
Lévi-Strauss, C. (1966). *The Savage Mind*. Chicago: University of Chicago Press.
Lévi-Strauss, C. (1969). *The Raw and the Cooked: Introduction to a Science of Mythology* (Vol. 1). New York: Harper & Row.
McGinn, B. (1996). The End of the World and the Beginning of Christendom. In M. Bull (Ed.), *Apocalypse Theory and the End of the World* (pp. 75–108). Oxford: Blackwell.
O'Leary, S. D. (1998). *Arguing the Apocalypse: A Theory of Millennial Rhetoric*. Oxford: Oxford University Press.
Popper, K. (1957). Philosophy of Science. In C. A. Mace (Ed.), *British Philosophy in the Mid-Century*. London: George Allen and Unwin.
Popper, K. (2013). *Realism and the Aim of Science: From the Postscript to the Logic of Scientific Discovery*. Abingdon: Routledge.

Rabkin, E. (1986). Introduction: Why Destroy the World. In E. Rabkin, M. Greenberg, & J. Olander (Eds.), *The End of the World* (pp. vii–vxv). Carbondale: Southern Illinois University Press.

Raglan, L. (1956). *The Hero: A Study in Tradition Myth and Drama*. Mineola: Dover Publications.

Rank, O. (2013). *The Myth of the Birth of Hero: A Psychological Interpretation of Mythology*. London: Read Books.

Rickles, D. (2014). History and Mythology. In *A Brief History of String Theory* (pp. 1–18). Berlin: Springer.

Stocking, G. W. (1974). The Shaping of American Anthropology 1883–1911. A Franz Boas Reader. New York, Basic Books.

Strauss, W., & Howe, N. (1997). *The Fourth Turning: An American Prophecy*. New York: Broadway Books.

Walliss, J., & Aston, J. (2011). Doomsday America: The Pessimistic Turn of Post-9/11 Apocalyptic Cinema. *The Journal of Religion and Popular Culture, 23*(1), 53–64.

Weisberg, J. (2008). *The Bush Tragedy*. New York: Random House Incorporated.

Wolfe, G. (1983). The Remaking of Zero: Beginning at the End. In E. Rabkin, M. Greenberg, & J. Olander (Eds.), *The End of the World* (pp. 1–20). Carbondale: Southern Illinois University Press.

CHAPTER 2

Interrogating on the Essence of the Zombie World

Introduction

In the first chapter, we delineated the conceptual grounds of apocalypse theory, as well as the importance of adopting ancient myths to understand the present. Humans are myth-makers by nature. The same can be said with respect to the horror movies and the cinema as promising anthropological fields. Some studies emphasize the possibility to analyze the plot of horror movies as a valid path toward a better understanding of society. The fact is that in the post-modern world, where faith is not the cornerstone of culture anymore, fear and hope mediate between the lay-citizens and uncertainness (Gasparini, 2015; Skoll & Korstanje, 2014). No doubt, the zombie world has notably resonated in the movie industries not only giving to a wider public a new form of entertainment but also speaking of us as the society (Birch-Bayley, 2012; Krzywinska, 2008).

As this backdrop, those discourses and narratives beyond the figure of living dead exhibit a fertile ground for cultural studies to infer in the roots of bottom-days. One might imagine that one of the most popular TV series *The Walking Dead* symbolizes a new phenomenon started with George Romero's horror film genre in the mid of 1970s. Romero's career does not need previous presentation since he was a pioneer in the imaginary of the undead and zombies. In 1968, he created *Night of the Living Dead* which is followed by *Dawn of the Dead* (1978), *The Day of the Dead* (1985) and a new remake of the *Dawn of the Dead* in 2004. In some perspective, his films, as well as many of the narrative of a post-apocalyptic world infested by Zombies, have the following features: (a) there are outbreaks of undead who invade the spaces of humans. These undead in various forms may not inhabit the earth; (b) by action of a virus or a mysterious event, zombies not only proliferate in the earth but also press human existence toward the disappearance; and (c) zombies, though the argument differs in each case, represent the decay of the modern forms of production and the liberal market. James Jameson argues convincingly that the post-modern culture is legitimated by the political unconsciousness, which means the lack of coherence or at the least causality between the acts and their effects. In the post-modern logic, there is nothing like the start of humanity, though at the bottom, the end sounds like a fascinating genre of consumption. To put the same in other terms:

> [It is a] 'degraded' landscape of schlock and kitsch, of TV series and Readers' Digest culture, of advertising and motels, of the late show and the grade-B Hollywood film, of so-called paraliterature with its airport paperback categories of the gothic and the romance, the popular biography, the murder mystery and science fiction or fantasy novel: materials they no longer simply 'quote', as a Joyce or a Mahler might have done, but incorporate into their very substance. (Jameson, 1991: 55)

Nobody knows where and when zombies will take appearance in this world, but what is clear, they are incompatible with human existence. In previous chapters, we discussed the role of technology in the acceleration of the moral decline, or the greed, which leads the known world to its own destruction. It is safe to say that the same observations apply to the zombie world. As Juliet Lauro and Karen Embry put it, the archetype of zombies surfaced in the mid of twentieth century as a new category which subverts the existent condition of mankind. The zombie lacks any conscience, while as a specter it evinces the cultural tension between the imperial power and the master-slave dialectic of the post-Haitian geography. For the Haitian tradition, the zombie is a creation of witchcraft which brings death to life. However, the importation to the US speaks of a monster who eats human flesh (or brains) interrogating on the individuality of the modern man. Starting from the premise that the zombie places our existence in jeopardy, Lauro and Embry contend eloquently that the psychological fear consists in imagining all we are potential zombies though our individuality struggles against fate. The survival harks back a temporal effort of resistance where death looms everywhere. This seems to be the essential cultural value of late capitalism where the exploited peoples remain unaware of their imprisonment. In the post-human condition, the dead and the life converge into a sublimated form that alienates the citizen into a fictional world.

Hence, the current chapter examines the metaphor of the zombie as a disembodied entity which interrogates furtherly on the human existence. As a result of a virus outbreak or at the best as the product of human greed, zombies strand in the world, threatening life as it is known. Based on the novel *World War Z* (authored by Max Brooks, a well-known novelist in the genre), we examine the main cultural elements behind the archetype of the *Judgment Day* constituted by the rise of zombies. There are some commonalities between our theory of Thana-capitalism and the fascination of global spectorship for the cannibalism of living dead. Survivalism, which is a residual value of Darwinism, plays a crucial role marking the destiny of those who are marked to be salved or condemned. Being in movement is the only criterion that distinguishes the real hero from the rest of mankind which is doomed to perish. In third, death (or in other terms, the Other's death) circulates as the main commodity of a society which is enrooted in solipsism and the end of all certainties.

The Literature and Cinema As Anthropological Fields

Although since immemorial times literature has played a leading role in not only educating but also entertaining citizens, it is equally true that literature exerts an ideological function, which merits the attention of scholars and critical thinkers. As M. L. Pratt (2007) brilliantly observed, literature in general and novels, in particular, are fertile grounds for social analysis. She widely showed how literature confirmed and reinforced the colonial rule regarding the presence of the alterity. The non-Western "Other" was portrayed as a subordinated (inferior) agent in respect to the Europeans. As a well-established genre, travel-writing historically legitimated the Western rationality, which is based on the assumptions that the New World should be systematically classified following the rules of science. In fact, as she remarks, the emergence of modern science and colonialism seems to be inextricably intertwined. Another seminal book, which strips literature from its sainthood, *Orientalism*—which was originally authored by Edward Said—exerts a radical criticism to the Academia reminding that far from disappearing, the dominant stereotypes forged during the colonial period are operating through novels and literature. The figure of *Orientalism* is not only biased but also singled out according to a previous cultural matrix that poses Europe as the best of feasible worlds (Said, 1995). The connection of literature and ideology was amply discussed in different contexts (Farrell, 1942; Hall, 1993; Irele, 1981), but what is more important, its efficacy lies in the possibilities to present the European mainstream cultural values as universal or applicable to other cultures. This type of ethnocentrism is replicated not only through literature but in cinema as well. K. G. Hall (1993) points out that the adoption of modern literature coincides with an ideological attempt that makes dependent the interests of lower classes to the ruling elite. Methodologically speaking, other scholars alert on the needs of focusing in novels and travel-writings to expand the current understanding of the society. Underpinned in the proposition that society can be understood by its prejudices and stereotypes, these voices strongly believe that literature expresses and extrapolates the much deeper values of society, as well as its hopes and anxieties (Karkuzhali & Raj, 2017; Lefevere, 2016; Raj, 2016). This conceptual position has been developed in post-colonial countries which have adopted a critical view of classic literature, as well as the Western democracy. The same account applies to the cinema industry. In this respect, the cultural anthropology has notably advanced toward the maturation of methodological instruments that help ethnographers to expand their understanding of cinema as a cultural expression of the society where it holds. As a form of ethnography, cinema provides professional fieldworkers with the necessary background to scrutinize a culture through visual ethnography. This argument was fleshed out by a new field known as "visual anthropology" (Griffiths, 2002). In a more than interesting work, *The Third Eye*, Fatimah Tobing Rony (1996) calls the attention how films serve as valid documents for the study of other cultures, but at the same time, he dangles that they may be the fictional platform of chauvinist discourses that subordinates other races to the whims of the white lords. The ethnographic cinema should be defined as an ongoing process of fieldwork which is a subfield of anthropology: "cinema is not only a technology, it is a social practice with conventions that profoundly shape its forms" (Rony, 1996: 8–9).

As cited above, Rony acknowledges that films allow the invention of a distant past which is re-elaborated or brought into the present. The cultural anthropology deciphers the complex codes of imperialism which are enrooted in the cinema industry. The propaganda that can be found in movies like *Tarzan* or *The Jungle Book* explains very well the positivist spirit where the non-Western "Other" is treated as inferior or uneducated. This happens because, after all, the cinema industry holds the power of molding minds combining surveillance and entertainment (Rony, 1996).

The velocity and accessibility of the audiences to multi-media content evolved successfully thanks to what Guy Debord termed as "the society of Spectacle" or John Urry "the tourist gaze". While Debord unfolds a Marxist critique on the society of consumption, where the concept of Spectacle is presented as the sign of moral decomposition, in Urry's studies, the emergence of the "gaze" structures a new stage of capitalism, where the material means of production are subordinated to a new aesthetic reflexibility. Tourists and travelers go there or here only to gaze what is unfamiliar to them. To some extent, they need novel gratifications and sensations which are encapsulated into a preceding cultural matrix (Urry, 1992). Rather, as per Debord, the logic of consumption imposes a new "representation" in which case the social ties are undermined in favor of the hyper-consumption and hedonism. The urgencies for maximizing pleasure can be understood as the correct dynamic of history, where the being declines before the having, and the having melts into the appearing. At the same time, the quality of life is bettered or the expectancy of life expanded, lay-people are ideologically ushered into a climate where the sense of reality is commoditized and circulated through the audiences. Per his viewpoint, the spectacle appears to be an inverted image of the society, where the image mediates and regulates the social relations (Debord, 2012). This raises the question to what extent the visual anthropology keeps a promising future in the days to come?

To answer this point, Professor Sarah Pink (2006) discusses in her book, *The Future of Visual Anthropology*, the complexity of media performance, associated to the potentialities of visual ethnography in an atmosphere, where the image dominates the demands of popular culture. The dominance of hypermedia presses ethnographers to adopt new tactics and methodologies, Pink says. From its onset, anthropology appealed to photograph and images as documents, but in the days of the global mass media, where events are immediately covered and disseminated to the spectatorship in seconds, anthropology faces serious challenges and dilemmas that need to be resolved. Paradoxically, though anthropology monopolized the idea of being-just-there watching and talking with the key informants, the inspection of movies and electronic screens were relegated as bit-players to the visual anthropology, a subfield of the discipline. For example, ethnographic photography was seen as a simple illustration instead of a reliable analytical toolkit in the 1970s. The irruption of phenomenology and the crisis it generates in the core of social sciences facilitated the arrival of visual anthropology as a serious (scientific) option. Phenomenology probed that two different fieldworkers may keep different positions, readings and interpretation of the same culture or the same tribe. In these terms, the methodological relativism questioned not only the ideals of an immutable sense of reality, which belongs to Enlightenment, but also the possibilities that history can be written and re-written according to the temporal interests of the dominant class. The reflexive approach, after all, comes to stay.

To cut the long story short, Pink argues convincingly that the rise of hypermedia needs form an interdisciplinary agenda, situating the process of reflexibility in sharp contradictions with other methodologies. To put this in other terms, while the observation of an image or a screen is the means through accessing the true, the process of reflexibility suggests the opposite, to be precise that the observed image mediates with the ethnographer's task. When this happens, the applied research leads toward to what Paul Virilio dubbed as the information bomb. From a tragic view of the world, he toys with the idea that the technological breakthroughs have changed the industry of transportation generating an excess of free time, which is filled by a hegemonic informational apparatus, which subverts reality into a fictional landscape (Virilio, 2005). Let's repeat readers that throughout a post-apocalyptic world, one image (news) sets the pace to a new one blurring the causalities between events, Virilio adheres.

SEEING IS BELIEVING

In a book that titles precisely like this subsection Seeing is believing, Peter Biskind (1983) envisaged the films as articulators of collective ethics saying what should be done or not. Through the narrative and plots imposed by films, audience embraces a necessary condition for social cohesion while keeping the loyalties on their respective leaders and institutions. Over decades, social sciences have questioned to what extent film-making and film audiences associate to the influence of ideology (Nichols, 1981; Ryan & Kellner, 1990). The theory of the hypodermic needle held that cinema industry applies an ideological pressure to generate indoctrination in the global audiences, though this argument was belatedly discarded in the passing of decades. Basically, each citizen actively engages and elaborates its own image of the film extrapolating not only its expectances but also its inner-world (Denzin, 1991). As the example of Latin America shows, sometimes cinema handles and channels critical voices and discourses oriented to confront the status quo beyond the social conformity. Inspired by Castro's revolution in Cuba and the rise of some minorities in the 1960s, Latin American directors saw in their films an opportunity for political militancy (Pick, 1993). This begs a more than interesting question, is cinema a type of cultural entertainment used to control the critical mind or a platform to stimulate the political militancy?

From the outset, this concern accompanied the early works of Karl Marx. In his seminal book, *The German Ideology*, he outlines that capitalism obscures the means of exploitation masking the sense of reality. Through different false representations (false consciousness), ideology does the job subordinating the workforce to the interests of the ruling elite (Marx, 1972). This position was criticized and complemented by Louis Althusser, whose attempts were oriented to show how ideology recreates an imaginary landscape which binds the peoples to an unreal condition of existence. In order for ideology to work properly, there should be conditions of reproduction which should be granted by the nation-state. While states develop repressive mechanisms to keep the control of society (repressive state apparatuses), some psychological dispositions known as "ideological state apparatuses (ISA)" are culturally imposed to dissuade workers that the capitalist ethos is the best of possible realities. This foundational structure dominates the material reproduction of capital, as well as labor power (Althusser, 2006). Not surprisingly, Marxism and Marxists are being extensively objected by two other academic weaves: postmodernism and post-structuralism. The former signals to the ideology as an obsolete cultural construction, whereas the latter emphasizes on the active role of agency in the configuration of countless cosmologies which are articulated in the same society (Friedberg, 1993; Stam & Miller, 2000). Not surprisingly, some critical scholars claim that neither postmodernism nor post-structuralism left a methodological vacuum because of the lack of a firm background, which serves to analyze the individual behavior. The question whether each agent develops a proper interpretation of the plot leads to thinking that there is no solid basis to infer a set of coherent or at the best universal variables. The post-modern condition, in David Harvey's terms, blurred the causality between causes and effects ushering the society into a solipsist atmosphere which resulted from an anarchic state of fragmentation (Harvey, 1989). In the cultural studies which usually center on the movie industry, the lack of coherent frames, Professor Hyangjin Lee adheres, is conducive to the influence of the status quo. He puts the example of North and South Koreas. Both nations adopted different rhetorical discourses elaborating their own definition and interpretation of politics. Doubtless, politics and film-making are inextricably intertwined. While North Korea sheds light on the contradictions and economic problems of capitalism as poverty and the material exclusion of some classes, South Korea alluded to the needs of freedom and democracy as key factors toward economic prosperity. Films reflect not only the feelings and shared emotions of society today, but also how the past is being projected in the present. For fieldworkers to grasp an all-encompassing model that helps them in understanding the studied society, films offer fertile soils in view of the socio-cultural contexts where the plot of the film takes the room. The author goes on to write that:

> The use of historical materials in the film serves a double purpose. A cinematic reconstruction of past society puts history from a new perspective, and by doing so, it provides insights into the present state. While this study investigates these two interrelated aspects of the historical film, it is particularly interested in the latter effect: an ideological message that the film conveys about present society through its image of the past. By examining the ways in which previous experiences are reinvented on the screen, we can discern the subtle and complex operations of contemporary ideologies in everyday life. (Lee, 2000: 3)

As the first point of entry in the debate, the above-cited excerpt shows two interesting things to be taken into consideration. On the one hand, the question of identity plays a crucial role as the tug of war of nation-state, which ideologically indoctrinates not only the grass-root movements of workers but also the lay-citizens. On the other hand, the rivalry between the two Koreas evinces that in spite of the efforts to agree on some consensus by the side of Koreans, the political division which shapes ideologies occupies a central position. Films are ideological instruments that structurally keep the order in both sides of the river. This does not mean that the subjects are passive entities who are unable to elaborate, contradict and even reject the status quo's message. As Slavoj Žižek (1989) brilliantly observes, the power of ideology does not lie in the message it instills, but indeed it says much for what it silences. This Althusserian viewpoint leads Žižek to acknowledge that the classic Marxist definition of ideology should be at the least reconsidered. This raises a thorny question revolving around the fact that cinema represents an ideological apparatus that conditions the autonomy of subjects. Far from being closed, the polemic still remains alive.

It is noteworthy that the growing interests in cinema industry by critics and cultural theorists have been notably triplicated over the recent years. As a complex form of entertainment, films follow the same dynamic of myths and mythical heroes. They describe or tell an atemporal story presenting a philosophical quandary and explaining how things should be solved. To put this in bluntly, plots say the spectatorship about what is good or bad. Miriam Hansen (1991) introduces an interesting debate revolving around the hands of feminism. Per her viewpoint, from its inception, cinema industry reproduced the societal stereotypes commoditizing women to the gaze and taste of a masculine audience. Based on Habermas' contributions, she holds eloquently that one might understand the films as the pleasure of domesticating women's bodies. As she observed, the question of cinema is not related to the gender injustices but to the fact that it accompanied the radical transformation of public and private spheres in the capitalist society. On this point, Hansen writes:

> Film spectatorship epitomized a tendency that strategies of advertising and consumer culture had been pursuing for decades: the stimulation of new needs and new desires through visual fascination. Besides turning visual fascination itself into a commodity, the cinema generated a meta-discourse of consumption (not unlike one of its antecedents, the nineteenth-century world fair), a phantasmagoric environment in which boundaries between looking and having were blurred. (Hansen, 1991: 85)

What is equally important, the American cinema mutated from the propaganda-oriented to dissuade new immigrants of the ideals of capitalism and democracy toward the apocalypse, which became a focal point of self-reflexibility.

Methodologically speaking, cinema studies have evolved as an object promising option over the recent years but borrowing paradigms generated in other already-established disciplines such as anthropology, economy and sociology. Robert Ray (1985) holds that the complexity of films is given by the fact that these industries are embedded in the history of technology, commercial forms, the audience, politics and of course economics. In this way, different approaches which ranged from psychoanalysis to Marxism without mentioning structuralism have dedicated time and efforts in the study of cinema. Understood as an ideological structure, Marxists saw in cinema the reification—or in other terms the dominant discourse—of the bourgeoisie. Rather, structuralism defines the process of film-making as the efforts to create real guiding myths which are programmed to debate much deeper philosophical dilemmas. Psychoanalysts acknowledge often that movies are dream-like constellations disposed to interrogate the social id. Ray reconstructs the history of Hollywood since 1930 to offer an all-encompassing model that explains the intersection of movies and politics. In a nutshell, cinema represents a complex set of codified myths which never escape to the ideological contamination imposed by the status quo. Movies replicate a previous paradigm which is culturally legitimated in much deeper myth. The American society was founded in the premise of exceptionalism that places the US beyond the scrutiny of an external entity. The movie is carefully drawn taking care of the foundational values of society, as Ray concludes.

> The American Cinema's version of this traditional mythology rested on two factors. First, Hollywood's power ... to produce a steady flow of variations provided the myth with the repetitive elaborations that it required to become convincing. Second, the audience's sense of American exceptionalism. (Ray, 1985: 56)

Ray contends that the American hero debates itself between two contrasting poles. On the one hand, a moral dilemma suggests that hero struggles to protect others or some particular community (the moral center). In this vein, there is an official hero who follows the law and the ethical mandate of society to make this world a better place. This hero struggles against the evil without asking anything in return. On the other hand, an outlaw hero, Ray adheres, enters in scenes maximizing their benefits while preserving their autonomy and quest for freedom (the center of interest). Both opposing figures are historically condensed in the American cinema, as well as the foundational background of the US. From its outset, American cinema reproduced and continued the same archetypical logics of literature, as Ray explained:

The nature of the film, with its discrete still pictured tricked into motion by projection, had always implied animation, but in classic Hollywood's versions, America's traditional mythology had seemed genuinely alive. Strangely, that mythology, which began in folk tales and literature before finding its home in the movies, now survives primarily in private houses. (Ray, 1985: 368)

Echoing Biskind (1983), one might infer that cinema and politics are inevitably entwined. In the US, the notion of security was constructed revolving around an external enemy which looms from the dark shadows to destroy—if not poisoning—the democrat institutions. Such an ideological message is widely extended and enrooted in the American cinema. To the communists as a major threat or the red scare, as Biskind eloquently reminds, two other villains have added: the police and the psychiatrist. A bunch of movies narrate the history of psychiatrists who are the real villain, while more than a dozen plots place in the cop a type of new psychopath who takes justice in their hands. Another interesting tension is given by the juxtaposition of the local and the national. The local hero sometimes plays a leading role in confronting external national forces which threaten the community's well-being. Villains keep the same characteristic, which associates to the fact they are moving, maximizing their interests no matter the costs for the group. Although conformity was ambiguous for American social imaginary, it is no less true that the archetype of villains is defined by their being prone to individualism in a society where the values of cooperation and competence daily collide. To put the same in other terms, heroes are moved by protecting the community to the extent they often run serious risks or are subject to great tribulations, while villains are in quest of their own gratification and conformity. This begs a more than an interesting question, is the zombie a dormant and repressed fear of individualism?

Biskind offers an alternative answer to this point. In fact, he writes:

All films, both centrist and radical, influenced the values of their audiences. The radical film tried to undermine or subvert mainstream values that the audience had absorbed from sources like school, work, family, media, church, and so on. Centrist films, on the contrary, reinforced these values. They encouraged viewers to conduct themselves in the manner society wished and discouraged them from doing what society abhorred. (Biskind, 1983: 163)

Whatever the case may be, the cinema industry—within the US—dealt historically with a great philosophical dilemma: the enemy within. Biskind analyzes a great range of plots from horror to action genre reminding two significant aspects. On a closer look, there are serious suspicions and fears about the internal enemy. Sometimes in the bodies of communists, aliens or even vampires, what lies in the fact that despite their evil nature, they look like us! Secondly, the Puritan ethic, which has historically characterized the American society, imposes the mandate that governments should be strong and weak to achieve successfully the collective goals. Strong because in this way, the state allows the social peace exerting the monopoly of force, while the weakness is a necessary condition to avoid totalitarianism (Biskind, 1983). The same was inferred by Professor Sam B. Girgus. Per his stance, The Hollywood Renaissance (in the literature and Cinema) is born on a dangerous tension between the democratic institutions that makes the civil life possible and the needs of heroism which legitimates messianic leaders. Based on the Western films, Girgus (1998) explains that directors such as Ford, Capra and Kazan were underpinned into the same idea of portraying America as a land of prosperity and opportunities while as a project to be imported to the (uncivilized) world. In the next section, we shall discuss a sociological answer to the figure of zombie and living dead.

THE SOCIOLOGY OF THE ZOMBIE

In her path-breaking book, *Purity and Danger*, Mary Douglas (2003) wrote that the concept of taboo should be culturally deciphered as an aversion to a finally ingrained object, which exhibits the collective fear manifested to the advance of contamination. The taboo reminds the importance to impose barriers in the contact with the divinity, and in so doing, the sentiment of anxiety this ritual represents. Through the articulation of different rituals, human groups tend to control and reduce the anxiety in order to avoid social disintegration. The dirtiness exemplifies the secondary result of the symbolic order, as well as the system of classification which identifies and expulses the undesired elements. In this way, the concept of purity enacts a social cohesion based on the influence and power of symbolism. Douglas recognizes that when the external events cannot be classified, the ambiguity emerges. The irrationality beyond the taboo not only legitimates the dominant rule but helps to understand such a traumatic experience. This conceptual backdrop is vital to discuss the origin and nature of living dead as well as the tendency of a modern audience to gaze at these types of products.

The zombie culture now represents more than 5 billion dollars per year, situating as a promising and growing cultural industry including video games, movies, comics and other cultural consumptions (Ogg, 2011; Platts, 2013). For sociologists and anthropologists, the field of zombies has received marginal attention. Instead, their interests were primarily given to witchcraft or religion. The current applied research feds back from the available ethnographies, fieldworkers' notes or documents left by scientists during the colonial period. No matter the ideological background, all these studies agree that zombies have evolved from folkloric traditions forged in the post-colonial landscapes toward an allegory of the worker's oppression (Dendle, 2007; Golub & Lane, 2015; Guynes-Vishniac, 2018; Lauro, 2017; Lauro & Embry, 2008; McIntosh, 2008; Puey, 2013).

Let us clarify that the term comes from Haitian *creole*, zonbi, which means "the lack of spirit". This fictional figure was originally introduced in the cinema industry through the production of Edward Halperin's (1932) *White Zombie*. This horror movie was based on the novel *The Magic Island* which sees the light of publicity in 1929 by the hands of William Seabrook (2016). One of the experts in the field, Kyle Bishop (2008) argues convincingly that the modern zombie should be geographically located to Haiti representing the dialectical relations between the European masters and their slaves. The archetype of the zombie exhibits a much deeper fear of being controlled by the local native in which case the imperial rule would be placed in jeopardy.

> The zombie was an ideological manifestation of the social and political superstructure in these newly liberated colonies, using fear to encourage hard work and subservience. When the western cinematic versions of these folkloric creatures are examined, zombies must be recognized as metaphorical manifestations of the Hegelian/slave relationship and the negative dichotomous social structure of colonialism. (Bishop, 2008: 145)

In part, the success of *White Zombie* speaks of the values of American society through the 1930s. The idea of zombification not only disposes of the Other's body but also confronts to the imperialist hegemony over the colonies. Zombies are commanded by their masters in the same ways slaves should be guided by Europeans. Centered on the rituals of voodoo, this zombie differs from Romero's versions. Here, the fear is strongly associated to be controlled and enslaved by the sorcerer than killed as in the contemporaneous horror movies. Zombies represent the under-class, the specter of those unrepresented who marginally struggle only to survive. In the capitalist system, the same can be said about the relation of the bourgeois class and the hapless workforce. To a closer look, for the Western imaginary, it is preferable to suffer an unjust death abroad than being enslaved beyond the borders of Western civilization (Bishop, 2008). This happens because the white supremacy consolidated in history on the basis of two main pillars. The first was the belief that Westerners brought a strong moral purity, which should be implanted in the world. Secondly and most important, the dichotomies between religion (rationality) and pagan rituals (irrationality), far from being closed, have certainly accompanied the dominant discourse of colonialism over centuries.

For Jean and John Comaroff (2002), the archetype of zombies speaks us of clearly the exploitative power of global capitalism which places the workforce into a senseless state of alienation. The Comaroffs interrogate on the dichotomies of liberal capitalism, which creates an ambiguous position. On the one hand, a ruling class controls the technologies of production leading to lay workers to pauperism and poverty. On the other hand, this situation accelerates the mass migration of workers to other places and cities. These migrants—like zombies—are in quest of opportunities though in an erratic way. The global capitalism dynamites the conditions of work creating "zombies" who strand elsewhere in quest of a better situation. In this way, Comaroff and Comaroff find that the economic crisis, the gift-exchange and the figure of culture in a global society of consumption are inevitably entwined. They go on to say:

> In the upshot, the two sides of millennial capitalism, post-apartheid style, come together: on one is the ever-more-distressing awareness of the absence of work, itself measured by the looming presence of the figure of the immigrant; on the other is the constantly reiterated suspicion, embodied in the zombie, that it is only by magical means, by consuming others, that people may enrich themselves in these perplexing times. (Comaroff & Comaroff, 2002: 792)

To cut the long story short, for the Comaroffs, zombies or the living dead embody the social anxieties and fears revolving around the figure of immigrants while, so to speak, those means of production that caused the pauperization of workforce, far from being corrected, are widely replicated.

As discussed, Peter Dendle (2007) dangles that zombies are something else than simple erratic monsters. Rather, they personify the different political and economic anxieties which are enrooted in the American culture. Expressing the same in two conceptual poles, these figures oscillate from narratives that describe the colonial oppression in the overseas territories (i.e., Haiti) to modern metaphors that speak of rank-and-file workers' exploitation. Emerging after the 1930s, the living dead adopted partially the features of Frankenstein and Dracula, which were encapsulated in the Western social imaginary.

Similar to the previous argument and continuing Dendle's debate, Todd Platts (2013) holds the thesis that zombies embody a natural fear linked to the loss of autonomy and final death. Today's zombies remind not only the importance of apocalyptic narratives but also our own impossibilities to cooperate with others. Platts starts from the premise that cultural productions evince something significant of the culture where they were created. A sociological explanation that answers why zombies matter suggests that zombies are real expressions (symptoms) of cultural anxiety, which mutates to countless forms ranging from nuclear war to unknown lethal viruses without mentioning terrorism and 9/11. In this way, each period produces specific-context fears and imaginings toward the consolidation of zombie culture.

It is important not to lose the sight of the fact that the living dead holds the key for a philosophical interrogation about life, death, liberty or autonomy and slavery, but what is more important, as a mythological narrative, it elucidates the contradictions of the capitalist system (Macfarlane, 2018; McAllister, 2012).

In earlier approaches, we laid the foundations toward a new understanding of zombie culture. While originally the archetype of zombies was imported from Haiti, where the US had exercised a military occupation, the entity adopts previous and much deeper assets which were present in Norse Mythology. The needs of devouring human flesh (cannibalism), associated to the acts of decapitation or shooting in their heads which are the only success ways of killing zombies, are present in the cult of fylgjur (the cult of fylgja). The Norse myths are commonly based on the figure of fylgja, a macabre soul derived from an enemy killed in the battleground that in the nights of winter devours human flesh. Like a living dead, the fylgja exhibits not only the duality of the soul but a type of the second id which remained even after death. To avoid the return of the fylgja, enemies were executed taking care of some rituals as decapitation or cremation. Once returned from death, these vengeful entities sowed the terror and desolation everywhere (Korstanje, 2009). Lastly, beyond its nature, the zombie culture has notably increased over the recent years, so to speak just after 9/11 and the turn of the century. Hubner, Leaning, and Manning (2015) tell us about "the zombie texts", which is a term coined to allude to those emerging narratives that posit not only the history of the zombie but also its psycho-philosophical constitution. To put the same in bluntly, zombies and living dead are social construes embedded in socio-economic and cross-cultural settings, which are historically determined.

WORLD WAR Z (A MAX BROOKS NOVEL)

This novel is doubtless inserted in what Hubner et al. (2015) termed as "bio-zombies", which means entities derived from the infection of a lethal unknown virus. The novel inspired the American apocalyptic horror movie in 2013 starring Brad Pitt, Dede Garner, Jeremy Kleiner and Ian Bryce among others. This book centers on an apocalyptic scenario where the human race is threatened by a strange virus, which, transmitted by a bite, infects and transforms healthy persons in zombies rapidly. Unlike other similar movies where zombies are slow moving, for this case, they are faster and stronger.

The plot starts with the narrator, Max Brooks (an agent of the United Nations), who describes an apocalyptic landscape just after the outbreak of a zombie virus. Apparently, a Chinese youth geographically located in the small town of Chongqing (China) is the zero-patient.

> There were seven of them, all on cots, all barely conscious. The villagers had moved them into their new communal meeting hall. The walls and floor were bare cement. The air was cold and damp. Of course, they're sick, I thought. I asked the villagers who had been taking care of these people. They said no one, it wasn't safe ... the villagers were clearly terrified. They cringed and whispered; some kept their distance and prayed. (Brooks, 2006: 6)

The virus rapidly spreads to other developing countries which operate in the dark sides of organ trade. Once the virus expanded, Israel leaves the Palestinian territories which are apparently infected to build a cordon sanitaire. The borders are closed to all those who are neither healthy Jews nor Palestinians. This type of exemplary center seems to be immune to zombies remaining as a promised land for those desperate people who flee from the urban cities. The American government not only fails to contain the virus but also introduces a placebo (Phalanx) to give citizens a sense of false security. At the time this conspiracy is revealed by a journalist, the panic seizes the society. In a post-apartheid atmosphere, South Africans implement a macabre plan known as Redecker plan, which consists setting sanctuaries in specific areas as forms of baits so that the undead distracts, permitting others located in safe zones to re-group and evacuate. Meanwhile, in the US, people escape to the mountains once they discover zombies are sensitive to extreme cold. After seven years, a conference in Hawaii invites the surviving leaders aboard USS Saratoga. In a hot debate, the representatives do not agree with any firm solution. The main superpowers such as the US, the UK, Russia and France plan their own attacks against the living dead but without any results and in some cases suffering higher costs. After the efforts to contain the virus, Pakistan and Iran destroy each other in a nuclear war. Finally, the War Z redraws the geopolitical map of the world. While Cuba adopts democracy as the main form of government, Tibet is liberated from communists and North Korea remains completely desolated. The fate of the UK is not clear in the novel, but the royal family flew to Ireland. The American army reinvented itself launching an offensive against zombies using different new weapons. Zombies can be stopped shooting or hammering their heads. The army develops a multi-task weapon that lobotomizes the undead. In three years, the government attempts to reconquer the contiguous US. Ultimately, the United Nations deploys a great military force to eliminate the remnant living dead in infested zones. Meanwhile, the peace is slowly restored; other viruses for which humans were immune are arising. Doubtless, this places—once again—the future of humankind in jeopardy.

To a closer look, the novel is narrated in the first person, like Max Brooks as the agent of the United Nations (original Commission Report). This instills a much deeper terror, blurring the borders between fiction and reality.

After further review, some assumptions should be made. In the first pages of the book, Brooks writes:

> The first outbreak I saw was in a remote village that officially had no name. The residents called it New Dachang, but this was more out of nostalgia than anything else. (Brooks, 2006: 4)

The above-cited excerpt signifies the terror that derived from the unsaid, from the unknown or in other terms from the uncertainty. Brooks' novel evinces an apocalyptic virus may very well surface anytime and anywhere beyond the borders of the civilized world. In this instance, China, aligned with Cuba and North Korea, is portrayed as an uncivilized nation (failed states) where for some cultural reasons the democracy never prospered. For Chinese peasant, the living dead were creatures, ghosts coming from the underworld. This irrational point is opposed to the rational role played by science in discovering the nature of the virus.

Secondly, the novel retreats how the government failed to implement successful programs of evacuation and contingency while focuses on corporate corruption as well as the existent tensions between Israel and Palestine in the Middle East. Brooks interrogates readers about the possibilities and the resiliency anyone should hold in emergencies like these. The culture of survivalism and the culture of preppers that have characterized Americans remain present in this book. Thirdly, in the post-apocalyptic world, the rise of living dead wakes up the civilization from its slumber. The excess of luxury and comfort leads toward a philosophical contradiction which needs to be reedemed by God. As stated in the first chapter, the mistrust in technology is a key element of bottom-day narrative, and of course, the zombie world seems not to be an exception. Brooks not only wanders through the world but also reports how it is gradually degraded. Once in Burlington, Vermont, he is nominated as Vice-president as a proof of courage but as a sign of moral decadence of professional politicians. Brooks in the battlefront struggled heroically to find a cure, while politicians deployed a set of unsuccessful steps to contain this crippling virus. He writes ironically:

> Elections? Honolulu was still a madhouse; soldiers, congressmen, refugees, all bumping into one another trying to find something to eat or a place to sleep or just to find out what the hell was going on. And that was paradise next to the mainland. The rocky line was just being established; everything west was a war zone. I explained to the president that we didn't have the energy or resources to do anything but fight for our very existence. (p. 147)

This excerpt reflects the Ray's dichotomy between the outlaw hero embodied by Brooks and all army soldiers who fiercely fight against zombies, and the officialdom, the leaders and politicians, who are in a safe refuge as the USS Saratoga. Last but not least, Brooks uses the term Total

War to symbolize the lack of food, resources and armies, as well as the impossibilities of militaries to fight against their comrades once they are killed, devoured and returned to life. What total war means for humans is brilliantly expressed in the following lines:

> All human armies need supplies, this army didn't. No food, no ammo, no fuel, not even water to drink or air to breathe! There were no logistic lines to sever, no depots to destroy. You couldn't just surround and starve them out, or let them wither on the vine. Lock hundreds of them in a room three years later they'll come out just as deadly. (p. 272)

After all, like terrorists, they never surrender, they never negotiate! This poses a serious dilemma with respect to the human capacity of trading. While the involving leaders debate what is the best course of action to follow, zombies move as a strong force that devours everything in its path. The politics is war by other means. Brooks reminds the philosophical dilemma of politics which is consolidated through the articulation of controlled violence. Wars signal to more than our greed, they are economic forms of production that shape our society, but what is more important, this war is pretty different. Mankind, after years of conflicts, is facing a new enemy who wishes our total annihilation as species.

> For the first time in history, we faced an enemy that was actively waging total war. They had no limits of endurance. They would never negotiate, never surrender. They would fight until the very end because, unlike us, every single one of them, every second of every day, was devoted to consuming all life on Earth. That's the kind of enemy that was waiting for us beyond the Rockies. That's the kind of war we had to fight. (p. 357)

The end of the novel is known for all readers who had come with this interesting text. It was not the goal of this chapter to describe or review the book but only giving some insight into the key elements of the dominant-discourse beyond the writer. An exemplary center symbolized by the US and Israel is confronted by China and Easterners. While the virus outbreak situated originally in the uncivilized world (China), West devotes efforts maximizing the existent instrument to protect humanity. It was unfortunate that zombies stay ahead in the war, but it interpellates the human character which is never brought to its knees. Humans always fight no matter what the conditions or the risks they should face.

In earlier approaches, we have proposed a new model to understand the modern cultural entertainment industry and the tendency for gazing the Other's suffering. From movies to video games, citizens feel certain happiness in consuming—through the media—disaster-related news or news containing mass-death and violence. The human suffering and mass-death situated as a criterion of attraction, something that worth the time to be gazed upon. The modern society, which sociologists such as Beck and Giddens imagined, sets the pace to a new (more macabre) world, dubbed as Thana-capitalism where death is the touchstone of society, the point that mediates between citizens and their institutions. The best example that illustrates how this world works is the novel *The Hunger Games* (Collins, 2008). Thirteen districts are oppressed by President Snow, who annually organizes some bloody games to cement his hegemony. The participants enter in the game exaggerating their possibilities and skills to win the game. Struggling in a Hobbesian climate of all against all, participants are killed one by one. The same happens in the liberal market where the big fish eats the small one. The Other's death gives further pleasure to participants who believe after all they are in trace toward the final game. Because they exaggerate their possibilities, they are unable to cooperate with others to outcast President Snow. The life in the Thana-capitalism can be compared to this dystopian scenario (Korstanje, 2016), but this point will be explained further in the conclusion.

Conclusion

After further discussion, the present chapter proffered a useful insight on the world of living dead. Anthropology, over the recent years, advanced in the study of novels and movies as valid methodological instruments to understand society within which they were created. The content of analysis is vital to grasp the discourses and narratives that speak us of our world, our cosmologies and fears. In the first section of the present chapter, we discussed the strengths and weaknesses of cinema as an anthropological field of analysis. However, scholars are far to agree some consensus whether cinema offers an ideological platform to dominate passively the global audiences or functions as a form of expression that is individually imagined and shaped. While the former stance takes its cue from Marxian approaches, considering movies as ideological artifacts that delineate the individual behavior, the latter signals to the individual capacity to construct actively its sensible world behind films. The biography of the subject selects different aspects of the dominant narrative giving some autonomy with respect to the ideology. The third section pivots an important step in examining the sociology of the living dead, a field which was overlooked by classic and post-modern sociologists. To wit, the history of zombies in the post-colonial arena is carefully scrutinized. We toy with the idea that zombies represent the fear of the white ruling elite to be ousted or controlled by the local "Other" (the native). Brooks' novel—entitled *World War Z*—provides us with a better understanding of the tension between the apocalypse and technology, the life and death, the mankind and the monstrosity. The plot is situated in a post-apocalyptic landscape, where zombies have invaded, devoured and destroyed the world as we know it. Mankind struggles at pains against this new (generation) of the undead. Based on the social Darwinism that characterizes Thana-capitalism, these zombies are a new step in the evolutionary chain that threatens humans as never before. However, since the human soul fights even at odds against all, not everything is lost. Brooks marks an interesting point revolving around the uncertainty, as well as the failure of bureaucracy to prevent the social change. In the Thana-capitalist society, where the war of all against all seems to be a criterion of distinction, social Darwinism is the dominant ideology. This means that the evolutionary aspects of classes are given by their intrinsic capacities to fight and survive (avoiding cooperation as a sign of inferiority). The Other's death becomes not only a criterion of entertainment but the opportunity to continue on the trace. The point is that this is a game where the winner takes all. In the labor market (like in *World War Z*), people struggle to get a job, to have the right to be exploited by the capital-owners. Their individualist behavior is determined by survivalism in a world where few concentrates the major portion of wealth, while the rest dies with nothing. Ideologically, *World War Z* reminds how the dominion of the human soul is certainly given by the passive acceptance the culture of survivalism offers.

It is not surprisingly intriguing the correlation between 9/11 and the consumption of zombie products. To set an example, AMC Latin America production, *Fear the Walking Dead* (conceived as the prequel of *Walking Dead*), presents Nick (a drug addict) laying in an abandoned house. Once he wakes up, he realizes his girlfriend behave erratically. She is sick but looks like us. She is a zombie, the zero-patient of a virus that will start the apocalypse in the earth. Nick and his family (Travis, Chris, Madison and Alicia) face an extreme panic when they discover how gradually people are infected by this virus and their neighbors convert in soulless monsters thirsty of blood. Australian sociologist, Luke Howie, says that one of the frightening aspects of terrorism is the lack of certainness with respect to where and how the attack will be perpetrated. After all, terrorists look and dress like us, they are humans—like zombies—whose intentions are unknown to us (Howie, 2012). In the popular wisdom, the jihadists are often portrayed as promising young people who are brilliant and native of the society they hate. Born in the US or Europe by first generation, they are educated thanks to the hospitality of Western civilization. Instead of their parents who suffered countless difficulties and tribulations, they live in an Eden where all their needs are met. Instead of enjoying this situation of grace, they are radicalized in "the Islam" and introduced to a mythical travel (preferably to Middle East) to be in contact with terrorist cells. Once returned, their minds have been altered forever. They sacrifice a promising future or career sublimating their life to kill others. In the same way, Nick kills himself starting a trip to nowhere! He unwittingly opens the doors for a radical transformation paving the ways for the next step in human evolution (Korstanje, 2019). In some way, terrorism is radically changing how the alterity is constructed not only in the cinema but also in the cultural entertainment industry. The zombie is an undesired "Other" whose right of hospitality is rejected by the hosting society (Korstanje, 2017). However, this is a much deeper seated issue which merits to be discussed in the next chapter.

References

Althusser, L. (2006). Ideology and Ideological State Apparatuses (Notes Towards an Investigation). *The Anthropology of the State: A Reader, 9*(1), 86–98.
Birch-Bayley, N. (2012). Terror in Horror Genres: The Global Media and the Millennial Zombie. *The Journal of Popular Culture, 45*(6), 1137–1151.
Bishop, K. (2008). The Sub-Subaltern Monster: Imperialist Hegemony and the Cinematic Voodoo Zombie. *The Journal of American Culture, 31*(2), 141–152.
Biskind, P. (1983). *Seeing Is Believing: How Hollywood Taught Us to Stop Worrying and Love the Fifties*. New York: An Owl Book.
Brooks, M. (2006). *World War Z: An Oral History of the Zombie War*. New York: Three River Press.
Collins, Z. (2008). *The Hunger Games*. New York: Scholastic Press.
Comaroff, J., & Comaroff, J. L. (2002). Alien-Nation: Zombies, Immigrants, and Millennial Capitalism. *The South Atlantic Quarterly, 101*(4), 779–805.
Debord, G. (2012). *Society of the Spectacle*. London: Bread and Circuses Publishing.

Dendle, P. (2007). The Zombie as Barometer of Cultural Anxiety. In N. Scott (Ed.), *Monsters and the Monstrous: Myths and Metaphors of Enduring Evil* (Vol. 38, pp. 45–57). Amsterdam: Rodopi.
Denzin, N. K. (1991). *Images of Postmodern Society: Social Theory and Contemporary Cinema* (Vol. 11). London: Sage.
Douglas, M. (2003). *Purity and Danger: An Analysis of Concepts of Pollution and Taboo*. Abingdon: Routledge.
Farrell, J. T. (1942). Literature and Ideology. *The English Journal, 31*(4), 261–273.
Friedberg, A. (1993). *Window Shopping: Cinema and the Postmodern*. Los Angeles: University of California Press.
Gasparini, S. (2015). *Zombis, fantasmas y la representación de la violencia en la narrativa argentina reciente (Zombies, Ghosts and the Representation of Modern Violence in Argentina)*. XXVII Jornadas de Investigacion del Instituto de Literatura hispanoamericana, Facultad de Filosofia UBA, Buenos Aires, Marzo.
Girgus, S. B. (1998). *Hollywood Renaissance: The Cinema of Democracy in the Era of Ford, Capra, and Kazan*. Cambridge: Cambridge University Press.
Golub, A., & Lane, C. (2015). Zombie Companies and Corporate Survivors. *Anthropology Now, 7*(2), 47–54.
Griffiths, A. (2002). *Wondrous Difference: Cinema, Anthropology, and Turn-of-the-Century Visual Culture*. New York: Columbia University Press.
Guynes-Vishniac, S. (2018). The Zombie and Its Metaphors. *American Quarterly, 70*(4), 903–912.
Hall, K. G. (1993). Literature and Ideology. In *The Exalted Heroine and the Triumph of Order* (pp. 3–15). London: Palgrave Macmillan.
Hansen, M. (1991). *Babel and Babylon: Spectatorship in American Silent Film*. Cambridge: Harvard University Press.
Harvey, D. (1989). *The Condition of Postmodernity* (Vol. 14). Oxford: Blackwell.
Howie, L. (2012). *Witnesses to Terror: Understanding the Meanings and Consequences of Terrorism*. Basingstoke: Palgrave Macmillan.
Hubner, L., Leaning, M., & Manning, P. (2015). Introduction. In L. Hubner, M. Leaning, & P. Manning (Eds.), *The Zombie Renaissance in Popular Culture* (pp. 3–14). New York: Palgrave Macmillan.
Irele, A. (1981). *The African Experience in Literature and Ideology* (p. 137). London: Heinemann.
Jameson, F. (1991). *Postmodernism, or the Cultural Logic of Late Capitalism*. Durham: Duke University Press.
Karkuzhali, P., & Raj, E. (2017). *Subalternity and Literature*. New Delhi: Author Press.
Korstanje, M. E. (2009). El culto a los muertos: la mitología nórdica antigua en el cine moderno (The Cult to the Dead: Norse Mythology and Modern Cinema). *Revista Chilena de antropología visual, 13*, 61–78.

Korstanje, M. E. (2016). *The Rise of Thana Capitalism and Tourism*. Abingdon: Routledge.
Korstanje, M. E. (2017). *Terrorism, Tourism and the End of Hospitality in the West*. New York: Palgrave Macmillan.
Korstanje, M. E. (2019). *The Challenges of Democracy in the War on Terror: The Liberal State Before the Advance of Terrorism*. Abingdon: Routledge.
Krzywinska, T. (2008). Zombies in Gamespace: Form, Context, and Meaning in Zombie-Based Video Games. In S. McIntosh & M. Leverette (Eds.), *Zombie Culture: Autopsies of the Living Dead* (pp. 153–168). New York: Scarecrow Press.
Lauro, S. J. (Ed.). (2017). *Zombie Theory: A Reader*. Indianapolis: University of Minnesota Press.
Lauro, S. J., & Embry, K. (2008). A Zombie Manifesto: The Nonhuman Condition in the Era of Advanced Capitalism. *boundary 2, 35*(1), 85–108.
Lee, H. (2000). *Contemporary Korean Cinema: Identity, Culture, Politics*. Manchester: Manchester University Press.
Lefevere, A. (2016). *Translation, Rewriting, and the Manipulation of Literary Fame*. London: Routledge.
Macfarlane, K. E. (2018). Zombies and the Viral Web. *Horror Studies, 9*(2), 231–247.
Marx, K. (1972). *The Marx-Engels Reader* (Vol. 4). New York: Norton.
McAllister, E. (2012). *Slaves, Cannibals and Infected Hyper-Whites: The Race and Religion of Zombies*. Division II Faculty Publication 115. Wesleyan University.
McIntosh, S. (2008). Evolution of the Zombie: The Monster That Keeps Coming Back. In S. McIntosh & M. Leverete (Eds.), *Zombie Culture: Autopsies of the Living Dead*. New York: Scarecrow Press.
Nichols, B. (1981). *Ideology and the Image: Social Representation in the Cinema and Other Media* (Vol. 256). Bloomington: Indiana University Press.
Ogg, J. C. (2011). Zombies Worth Over $5 Billion to Economy. *24/7 Wall St, 25*.
Pick, Z. M. (1993). *The New Latin American Cinema: A Continental Project*. Austin: University of Texas Press.
Pink, S. (2006). *The Future of Visual Anthropology: Engaging the Senses*. Abingdon: Routledge.
Platts, T. K. (2013). Locating Zombies in the Sociology of Popular Culture. *Sociology Compass, 7*(7), 547–560.
Pratt, M. L. (2007). *Imperial Eyes: Travel Writing and Transculturation*. Abingdon: Routledge.
Puey, L. C. (2013). El mito del zombi en la actualidad: desmembramiento sacrificial colectivo (The Mythical Archetype of Zombie Today: Disemboding of the Sacrificial Collective Action). *Arbor, 189*(764), 089.
Raj, E. (2016). *Literature and Society: Challenges and Prospects*. New Delhi: AuthorPress.

Ray, R. B. (1985). *A Certain Tendency of the Hollywood Cinema: 1930–1980*. Princeton: Princeton University Press.

Romero, G. (1968). *Night of the Living Dead*. Image Ten. 96 Minutes. English.

Romero, G. (1985). *Day of the Dead*. Dead Films Inc. 100 Minutes. English.

Rony, F. T. (1996). *The Third Eye: Race, Cinema, and Ethnographic Spectacle*. Durham: Duke University Press.

Ryan, M., & Kellner, D. (1990). *Camera Politica: The Politics and Ideology of Contemporary Hollywood Film* (Vol. 604). Bloomington: Indiana University Press.

Said, E. W. (1995). *Orientalism: Western Conceptions of the Orient*. Harmondsworth: Penguin.

Seabrook, W. (2016). *The Magic Island*. Mineola: Dover.

Skoll, G., & Korstanje, M. (2014). The Walking Dead and Bottom Days. *Antrocom: Online Journal of Anthropology, 10*(1), 11–23.

Stam, R., & Miller, T. (2000). *Film and Theory: An Anthology*. New York: Wiley-Blackwell.

Urry, J. (1992). The Tourist Gaze "Revisited". *American Behavioral Scientist, 36*(2), 172–186.

Virilio, P. (2005). *The Information Bomb* (Vol. 10). London: Verso.

White Zombie. (1932). Victor Halperin. US, English, Halperin Productions.

Žižek, S. (1989). *The Sublime Object of Ideology*. London: Verso.

CHAPTER 3

The Undesired Other

Introduction

Ethnocentrism is defined as an ideological closed system that prejudges others cultures according to the mainstream cultural values of the in-group (Andersen & Taylor, 2007). As Marvin Harris eloquently evinced, European nations launched to colonize vast overseas territories with the end of indexing new economies while reaffirming their cultural supremacies over others natives. The Western paternalism sustained the fact that aboriginal cultures were frozen in a premodern state similar to what Europe faced in ancient times. The modernization, the free trade as well as the technological breakthrough ushered in a new stage of production and evolution that crystalized ultimately in modern Europe. While the European colonial rule expands worldwide, many aboriginal and native cultures run the risk of disappearance. As a result of this, scientists and above all anthropologists have the mandate to study these cultures before their exhaustion. Harris argues convincingly though anthropology was not conducive to the exploitation natives suffered—it was inevitably entwined to colonialism. The European ethnocentrism was based on a strange paternalism. While it moves toward the needs of exporting the European values to the world, the advance of the material progress paves the ways for the disappearance of entire cultures. On one hand, natives were seen (under the logic of noble savage) as essentially well, but on the other hand, they were irrational or illiterate. Europeans, who dominated the arts and sciences, have moved forward the evolutionary pyramid but sacrificed their own freedom (Harris, 2001).

Even if colonialism has ended after WWII, the old logic still prevails. The British and French empires succumbed, while the US emerged as successor of European civilization. In this vein, the 9/11 and the resulting War on Terror reinforced some old dormant prejudices and ideas which coined by colonialism were fagocitated by political analysts during the Cold War. The figure of the rogue state was accompanied by the rise of presidential speeches—like Clinton in 1994 or Bush declaring the war against the Axis of Evil—that encouraged the US and its allies to protect democracy worldwide. In this chapter, we discuss critically to what extent our idea of democracy is enrooted in sentiment ethnocentrism which claims that democratic governments are superior to undemocratic nations. The figure of the Other emulates and interrogates us. Our thesis is that terrorism not only decomposes and destroys the social ties, but also is closing the openness to the difference that characterized the Western civilization. We navigate through the waves of polemic waters ranging from abortion to torture (without mentioning the questioned American exceptionalism).

This chapter deals with not only the problem of "the undesired Other", but the conception of feminism, as an ideological, academic and political movement marked to confront the patriarchal order toward equality of both sexes. In fact, feminism campaigns for women's right including domestic violence, social integration, fairer wages and reproductive rights (legal abortion). In 2018, though the Argentine Chamber of Deputies debated and ultimately passed the legal abortion, it was rejected by the Senate. Emulating the discourse of feminism in other countries, the movement promoted "legal abortion" under the lemma "mi cuerpo, mi decision!" (my body, my decision). Beyond the reasons that lead women to end her pregnancy (which is very personal), we review in depth the cultural tendency to neglect and control—if not eradicate—"the Other" when it is not desired. Like zombies, these undesired Other may be expatriates, migrants, terrorists and babies. As a product of the fear instilled by terrorism, West is facing the erosion of its cultural background, even the hospitality.

The American Exceptionalism

In recent years, some specialists have energetically questioned the role of the US as the watchdog of human rights and democracy throughout the globe. To some extent, the US vindicates the respect for the individual rights in other autonomous nations, while belligerent accusations on the ways prisoners are treated in Supermax prisons or other terrorist detention camps at Guantanamo or Abu Ghraib are ignored (Korstanje, 2019; O'Brien, 2004; Roth, 2006; Skoll, 2016). The question whether torture was not accepted by ethicist scholars was indifferent to American citizenry, which prioritized the security to the human rights (Gronke et al., 2010). Paragraphing Hobbes, one of the dilemmas of liberal state seems to be how to resolve the individual freedom and the monopoly of violence (Hobbes, 1968).

The controversy led to the liberal scholar Michael Ignatieff holding that torture is useful if it can be duly regulated by the democratic institutions. For Ignatieff, torture is esteemed as the lesser evil before the advance of terrorism in the world (Ignatieff, 2013). The same applies to the protocols and programs nations promote to deter the contamination and the greenhouses gases that contribute to the climate change. The US keeps a double moral standard. While internally the state encourages the economic prosperity, the free trade which supposedly improves the functioning of democratic institutions externally imposes "extractive institutions" which are oriented to repatriate basic resources to the homeland metropolis (Clark, 2005). In his book, *The Challenges of Democracy in the War on Terror*, Korstanje (2019) affirms that emulating the tactics of conquest of the UK, which combined an internal climate of freedom with oppression and violence at the periphery, the US historically developed an imperial system orchestrated under the lemma, divide to rule. The different American governments not only adopted a negative viewpoint of the non-Western "Other" but also feared the advance of non-democratic regimes (Korstanje, 2019). Hence, Americans were certainly entrapped in an atmosphere of narcissism that presented them as special, different and invested by the divine mandate to export democracy and capitalism as the best of feasible worlds. The American expansionism, doubtless, was historically associated with the idea that there are "rogue states" or "failed states" which someday will compromise the security of the US. This was what experts dubbed as "American exceptionalism" (Ignatieff, 2009; Korstanje, 2016b; Lipset, 1997; Madsen, 1998; Skoll & Korstanje, 2013; Tyrrell, 1991).

There is no doubt at all that Max Weber not only was a pioneer in finding the connection of religion with the economy but he realized how the cosmology of Protestantism (unlike Catholicism) constructed a closed view of future, which leads gradually to exceptionalism. The introduction of predestination started an economy of salvation where its immediate consequences paved the ways for the rise of capitalism (Weber, 1958, 2013). Protestants feel and should demonstrate they are special because this is the only way to know if one pertains to the privileged group of chosen peoples. One of his contributions rested on a much deeper analysis of labor organization as well as how it determined the territorialization according to cultural archetypes. In a seminal book, entitled *Visionary Compacts*, Donald Pease (1987) dissects the tension between patriots and their doubts after Revolution and Great Britain. From its onset, the US debated on the dichotomy of freedom and the tyranny imposed by the British Crown. The level of anxieties produced by the separation between England and the Civil War not only constituted serious concerns for settlers but also created a gap between the idea of national unity and local cultures. While the government unified a central authority to subordinate the loyalty of citizens, it is undermining (like British Empire) the autonomy of peoples. One of the greatest fears of founding parents was based on the possibility to repeat the British Project. To shorten such a bridge, the discourse of manifest destiny gave to Americans a reason to trust in their leadership style. In the same way, the US did the best to forget the past, entering in contradiction with its own identity. The sense of manifest destiny not only resolved the long-simmering conflict of a new-born nation with its empire but also posed the hopes that democracy should be imported and emulated to the rest of nations. As a result of this, US coined a new paternalism which accompanied their ethnocentrism up to date. In this vein, S. Coleman (2013) held the thesis that American fundamentalism orchestrated the essence of religion with politics, alerting that the world is always a dangerous place. These new orders seem to be charged with the needs of reforming the world, expiating their sins by means of sacrifice and a renovated atmosphere where fear and grace converge. Americans and other Anglophones, especially those in Britain and the settler countries, Australia and Canada, have produced a culture of terror. That culture induces a generalized fear among the populations of those countries. With a focus on the US, the ruling class has constructed a culture of fear that has evolved from the kind of fear associated with the anti-communist hysteria in the years following WWII and its predecessor Red scares to its current incarnation of the terrorism obsession (Korstanje, 2019). While recognizing popular participation in constructing this culture of fear, the fact is that elites in the centers of world capitalism have fostered its construction with planning and deliberation. The culture of fear is conducive in keeping class conflict in America and the world under control (Skoll, 2016). In sharp contrast with Spanish settlers, English colonization crystalized by the use and abuse of trade. The English reserved their right of intervening in the autonomy of natives, organizing their life but excluding them from their original project. Their success in expanding a subtle but stronger mechanism of discipline was based on the declaration of White supremacy, which led to racial connotations (Guidotti-Hernández, 2011). As Richard Hofstadter eloquently remarked, this sentiment of exemplarity emasculated a dormant discourse of racial supremacism which accelerated the adoption of social Darwinism as the main ideology of the country. Invested in the rights to command the destiny of the World, Anglo-Saxons not only expanded their ideology to the periphery, during the nineteenth century, they reformed the original Darwinian thesis on "the survival of the fittest" as the "survival of the strongest". This suggests that becoming a central power gives its own right to reinforce peculiar ethnocentrism, which is legitimated through the manipulation of future. The idea of controlling future gives legitimacy to the political powers. Hofstadter is not wrong when he says that the only primary aspect to rationalize the competition among citizens was the natural selection proposed by Darwin (Hofstadter, 1992). In the text of authoritative voices as William Graham Sumner and Herbert Spencer, social Darwinism, far from being the theory created by Darwin, fits with the postulation of new emergent exceptionalism, which derived from Puritan tradition (FitzGerald, 1986). This raises a more than interesting point: what are the concrete effects of social Darwinism in capitalism, or to what extent this biological theory is conducive to capitalist class formation?

In fact, the introduction of social Darwinism was resisted by Religious leaders, until they realized that millionaires were not the result of greed, or the sin, as puritan claimed, but the consequence of natural selection and moral virtue. Millionaires and capital-owners were the best specimens of humankind simply because they were carefully selected by their strength, validated by their skills in business or abilities to foster competitive climates. Under an aristocratic view of the Republic, social Darwinists focused on the needs of recycling the society to cultivate elevated and reified citizens. In parallel, as Miller (1953) has widely discussed, the hostility of environment or neighbors (in this case the natives) for the Puritan mind represented the token they were in the correct side, sublimating their sins in faith.

As this backdrop, Calvinism adopted the notion of predestination to divide the world into two parts, those (selected peoples) who are allowed to enter in heaven, and those (expressed in the majority of peoples) who are doomed to hell. At a closer look, social Darwinism and Calvinism not only were inevitably entwined but fostered similar ideas with respect to the problem of identity. The social Darwinism was annealed to the underlying Calvinist doctrine of hard of individual salvation, stewardship and prosperity as a sign of moral superiority. Equally important, in earlier days, the civil war nativism surged a disciplining force that evinced resistance to new-comers (immigrants). Although originally nativism was merged with a dormant sentiment of anti-Catholicism, it is equally true that nativism was mutated to other forms in the south where slavery was finally rationalized and legitimated. The racial discrimination, as well as its practices, constructed a barrier between the community and these undesired guests (Janiewski, 1991). Underneath it all lays economic exploitation of new European immigrants and African American slaves. At the same time, the US carried out its long-term genocide of the North American Indians. Eric Cheyfitz explained that empires construct a subordinated image of the Other, who can never be equal to the elite. Ranging from ridiculing to demonization, the others are often portrayed as inferior, or uncivilized. Imperial discourse consists in disciplining this Other—African American slaves and their descendants, recent European immigrants and native North Americans—to make them decent citizens (Cheyfitz, 1993; Stilz, 2009). In practice, that came to mean becoming White (Ignatiev, 1995). From that moment on, the politics of exception, as it was studied in different fields, not only accompanied the US but also ignited a hot debate worldwide. As Korstanje puts it, this principle of exceptionality, which is proper of Americans, denotes a repressed sentiment of insecurity inasmuch as other cultures are suspected of being "rogue" or simply "enemies of democracy". Underpinned on the proposition that "pre-emption" was an adequate response to the rise of "evil-doers", "rogue states" or even "virulent terrorist groups", the US historically vulnerated the autonomy of others nations, under the auspices of the principle of self-representation, a right given by the constitution and the legal jurisprudence. In fact, the idea of American exceptionalism is directly enframed into the symbolic core of Protestantism and its legacy (Korstanje, 2013, 2016a).

Terror and Humanitarian Reasons

Terrorism is defined not only by the articulation of extreme violence against innocent civilians but also by the introduction of terror to discourage the political militancy of lay-citizens (Feierstein, 2017; Korstanje,

2016a; Timmermann, 2014). As Corey Robin (2004) brilliantly noted, over centuries political philosophers thought politics and fear as distinguishable and opposed categories, but at a closer look, they are inevitably entwined. The ruling class adopts the idea of an external enemy to indoctrinate their own citizens, while the internal element dissuades lay-people not to confront directly against the government (Robin, 2004). Professor Geoffrey Skoll from State University of New York (SUNY) at Buffalo holds the thesis that the process of globalization expanded and ensured by the US accelerated long-dormant racism toward the non-Westerners. The culture of fear permitted Bush's administration the intervention in economic resources otherwise would remain inexpugnable. For Skoll, the 9/11 allowed economic policies otherwise would be rejected (Skoll, 2016). Some critical analysts agreed that the fear of terrorism, as well as the rise of Islamophobia in the West, paved the ways for the reenactment of racial and xenophobic narratives that placed Donald Trump in the presidency of the US (Altheide, 2017; Chomsky, 2015; Puar, 2017). In this respect, terrorism has brought direct consequences, so to speak, in the economy of the US (Howie, 2007, 2017), while other indirect effects were overlooked by the specialists such as racist manifestations or hostilities against Muslim population (Abbas, 2012; Poole, 2002; Saaed, 2005), censorship for those scholars who do not agree with the geopolitics of the government (Doumani, 2006; Post, 2006) and travel ban for those who come from Middle East or nations historically related to terrorism (Altheide, 2017). In the book *Liquid Surveillance*, David Lyon and Zygmunt Bauman (2013) bring the problem of security into the foreground. They coined the term adiaforization to denote the obsession for military goals at any cost. In the War on Terror, the society ascribes to an end-justifies-means logic where civil victims are treated as collateral damage. In the liquid society, theorists contend, security, as well as the derivative surveillance technology, plays an ambiguous role. It is safe to say that it signals the needs of protection from external risks or enemies while marks the protected family as an exemplary unit which takes part of chosen peoples. Lyon and Bauman remind two important things to have into consideration in this debate. On the one hand, only a few portions of people in the world are legally allowed to travel. On the other hand, terrorism accentuates the material asymmetries produced by capitalism.

To some extent, the tight surveillance orchestrated by the US government at the borderlands and border checkpoints speaks not only of a sentiment of fear with respect to the alterity but also that after all terrorism won. In a book which has no reference in English, the French anthropologist

Didier Fassin (2018) argues that global capitalism rests on two main contradictions. While first world tourists are legally enabled to travel everywhere, migrants and exiles are doomed to live entrapped in a territory of poverty and pauperization conforming an underclass, which interrogates the essence of liberalism today. Echoing Foucault, Fassin alerts that one of the limitations of bio-politics consists in thinking the body as the result of the discipline which derives from needs of administering the life. Imposing politics oriented to expand life means the instauration of mass-death, as Fassin recognizes. Still further, if we start from the premise that the citizen embodies the power of the rule, it is no less true that today "the suffering" has situated as the mark from where "this under-class" demands for rights and protection. Exiles, unemployed citizens and refugees ask the developed nations for hospitality and shelter. Marxist theory speaks us of a society structured in classes, where the ruling group exploited the workers. This exploitation was covered through the articulation of different ideological artifacts. In this new global capitalism, the disciplined bodies exhibit their miseries (so to speak the marks) to captivate the attention of the public official. Within these horizons and constellations, suffering mediates between people and the democratic institutions. Fassin coins the term "bio-political legitimacy" to denote the relations of the suffering body and the nation-state. In view of this, three elements are vital in the configuration of a dominant narrative aimed at protecting the weaker: (a) the need, (b) the compassion and (c) the merit. The need appeals to the impossible condition of living in a place where the ontological security is threatened. The hapless guest requires and needs the assistance of the host state reminding an urgent situation otherwise means certain death. This opens the doors to the empathy that leads toward compassion. The public official selects carefully the biographies and stories among thousands of migrants enabling a circuit of gifts, where the pain is the main commodity. Ultimately, the figure merit signals to the fact that the applicant for asylum should show he or she deserves the help. This probably is the most significant aspect of modern hospitality because the applicant went through the medical and Western reason, evincing their wishes to get ahead. Paradoxically, even if accepted, the guest is subject to a much deeper instrumental reason. What is more important, the law divides the desired from the undesired guests while, in this way, its application is based on the negation of the Other. Fassin puts the example of Kosovo War when the American airplanes exceeded their flying altitude to avoid the enemy artillery (known as the doctrine of zero-casualty warfare). In consequence, their targets were more diffuse, causing hundreds of collateral damages and victims. This moot point reminds the idea that the bio-politics tended to protect a few selected group is successfully orchestrated through the exploitation and the inevitable death of the rest (Fassin, 2018).

What Is Wrong with Feminism?

In the Western nations, feminist movements struggled in order for women to gain further rights which include the right to vote, legislation regarding equal pay, social integration and domestic violence (Hawkesworth, 2006). Although originally inspired in the feminist theory, the movement advanced a lot in the reproductive rights including legal abortion and access to contraceptives (Messer-Davidow, 2002). From its outset, this political movement reached four well-differentiated stages. The first ranges from the nineteenth to the early twentieth centuries. At this point, activists fought for the promotion of material benefits for women who were historically relegated in the labor market. These benefits consisted of better salaries, equal contract, marriage and other property rights (Freedman, 2003). The second-wave feminism was born just after the mid-twentieth century in some European nations. Theorists of this movement find that the oppression is determined by cultural and political inequalities that led women to internalize the so-called supremacy of men (Dooling, 2005). The third- and fourth-wave feminism surfaced in the late twentieth and twenty-first centuries. Unlike the others, two previous schools, activists adopt a post-structuralist interpretation of politics, which not only is more radicalized but also leads toward a new understanding of the matrix of domination from where the patriarchal order operates (Leslie & Drake, 1997).

Because of limitations of space and time, we are unable to review in this chapter all the specialized literature which focused on feminism as the primary object of study. Anyway, *The Grounding of Modern Feminism*, an investigation presented by the American historian Nancy F. Cott, evinces a fertile source to debate the origin and evolution of feminism in the West. Although feminism is always a hard concept to grasp, as Cott adds, it seems equally true that the movement was originally associated to give further rights to women.

> Feminism posits that women perceive themselves not only as biological sex (but perhaps even more importantly) as a social grouping. Related to that

understanding is some level of identification with the group called women, some awareness that one's experience reflects and affects the whole. The conviction that women's socially constructed position situates us on shared ground enables the consciousness and the community of action among women to impel change. (Cott, 1987: 5)

Most probably, the movement rests on a philosophical (if not epistemological) paradox. While seeking unity, consciousness and equality, feminists do not escape to differentiate themselves from men. For Cott, feminism involves three interlinked arenas. The first was associated with the charitable goals of assistance and support to women by reason of their gender. The second was the activism in the post of those campaigns initialized for oppressed women to have further rights (e.g., to the universal suffrage). The third arena signals to more broad-ranging objectives aimed at understanding not only the dominant discourse but also those conventions, stereotypes and structures historically cemented by the law. Cott toys with the belief that feminism evinces one of the most difficult aspects of political liberalism: the inequalities among citizens within the US. Sooner than later, the claims of feminists were echoed by "blacks", rank-and-file workers and other minorities oppressed in the name of liberalism.

> The underlying theme was variable human beings as men were, had the same human intellectual and spiritual endowment as men, and therefore deserved the same opportunities and rights to advance and develop themselves, persistently surfaced. (p. 19)

Adjoined to the political opposition to the arbitrariness of the patriarchal order, the feminist parties struggled not only in improving their position but also in forming a shared discourse with respect to men. As Cott puts it, once feminism appeared, the gender assignment and the family unit was radically changed. While some voices alert women wanted to put themselves over men creating a war against sexes, Cott acknowledges feminism transformed the woman's agencies leaving the paradox for the next generation. How being human beings and women at the same time?

In a seminal book *The Science Question of Feminism*, Sandra Harding reviews and questions the epistemology of feminism as a cultural project. Per her viewpoint, the act of replacing men with women is far from the egalitarian sense feminism predicates. In fact, unless the rules and societal background that generated the asymmetries among classes and sexes are changed, feminism is short-lived. As she puts it:

> Feminism proposes that there are no contemporary humans who escape gendering; contrary to traditional belief; men do not. It argues that masculinity—far from being the ideal of members of our species—is at least as far from the paradigmatically admirable as it has claimed feminity to be. Feminism also asserts that gender is a fundamental category within which meaning and value are assigned to everything in the world, a way of organizing human social relations ... all that stands between us and that project are inadequate theories of gender, the dogma of empiricism, and a good deal of political struggle. (Harding, 1986: 57)

The concept of science was originally thought according to the binomials between rationality and emotions. While the former was epitomized in the same of men, the latter does for women. Science, for feminists, is structured in a set of value-laden false beliefs created to protect the interests of the patriarchy. Once the produced knowledge is decoded, fieldworkers grasp the ideas, narratives and the cultural values laying in the masculine domination. What Harding reminds seems to be the following axiom. Women were historically relegated by men from the productive cycles and spaces. The ideology, which was drawn by scientific research, controlled women for centuries, but it raises a thorny question: why do feminists think whether women take the control of the labor market they will not impose the same ideology?

> The emergence of masculine domination among our distant ancestors can be understood as the transfer of the conceptualization and control of women's sexuality, reproduction, and production labor to men—a process intensified and systematized in a new way during the last three centuries in the West. Here, too, the attribution of different natures and worldviews to women and men presumably occurs originally as an ideological construct by the dominators, but subsequently becomes true as the control of women's labor is shifted from women to men. (Harding, 1986: 188)

Last but not least, Anthony Giddens traces back the feminist movement to the industrial ethos. Per his viewpoint, sexes are social construes determined by the organization of labor, as well as the reproductive forces of society. The position of women, in vulnerable conditions, has notably benefited in the passing of years, but what is more important, the precaritization of men and women in the liberal market has been increased. Today women and men should compete in egalitarian conditions for a job, in which case, their identities have been radically changed. This happens because the liberal capitalism placed women in the same position as that of men. Far from being a positive thing, the emancipation of women represents the precaritization of work-force, and the reconfiguration of a new sexuality. The technological breakthroughs disposed to expand the life expectancy led women into a new organic sex, which invariably resulted in violent reaction by the side of men (Giddens, 2013).

The Abortion and the End of Hospitality

From diverse angles, this section discusses the question of abortion, as the direct result of terrorism, as well as the culture of fear. We hold that the modern self shows some limitations to understand and create empathy with the alterity. The right of legal abortion, as well as radicalized discourses in the west against the forced migrants, frames as a sign of the irreversible decline of hospitality.

As stated in the introduction, July and August of 2018 witnessed a hot debate in Argentina revolving around "the legal abortion", which facilitated for poor women the access to end a pregnancy at their discretion earlier than the three months of gestation. Although this bill was finally rejected by the Senate, the Catholic Church widely criticized the initiative calling for political militancy (crushing directly with feminists and Marxist parties). Quite aside from this, the debate was mediatically shaped involving the participation of notable people of Argentine culture as well as scientists, physicians and lawyers. The fact was that the medical reason, here, was opposed to the celebrities. As Žižek eloquently explained, abortion is an extreme measure that probably is practiced by higher and mid classes, though, in essence, they speak in the name of poor women. The feminist discourse precaritizes and stigmatizes the poor women imposing the ideals and desires of middle classes (Žižek, 2014). The power of hegemony consists in making others believe they are free, though essentially they behave or act in a way the ruling elite want. What is the intersection of abortion and hospitality?

In the earlier chapter, we coined the term "Thana-capitalism" to denote a new stage of production, where death is the main commodity. The culture of Thana-capitalism prioritizes the individual decision affecting the social ties which are necessary for social cohesion. In view of this, a

Darwinist climate of existence is circularly imposed (Korstanje, 2016b). Like zombies, rank-and-file workers struggle others workers for a job, while narcissism has empowered the cultural values of society (Lasch, 1979). The Other's suffering becomes a criterion of attraction and something that deserves to be gazed in the Thana-capitalist society. No matter than the part of the world, spectators consume death as a form of media entertainment. Everywhere in movies, novels, TV programs and documentaries, competition and death are two key elements of Thana-capitalism. The sacrifice is seen as a sign of weakness in a society where the winner takes all. The proper pleasure maximization is given by the Other's pain. This solipsist behavior is previously conditioned by an individual character that fears the alterity. In a more than interesting and conniving book, David Altheide alludes to the culture of fear, which is formed not only by the rise and expansion of terrorism but also by the mistrust in non-Western Other. Over the recent decades, terrorism eroded the basis of social trust allowing the emergence of radicalized voices (like Donald Trump) which manipulated and directed the fear instilled by terrorism toward the strangers. Terrorism is gradually changing the democratic institutions of the US, placing democracy in jeopardy (Altheide, 2017). Is abortion part of the same tendency?

At a closer look, legal abortion was legally sanctioned in developed nations, leaving the theme in a stalemate in Africa and Latin America. While some voices claim the Catholic Church played a leading role in preventing the legal abortion in Latin America, at a first glimpse, another interpretation suggests that the theme is placed in the agenda just after the 2000s in the hands of left-wing scholars, feminists and activists. This involves nations as Germany, Spain and Portugal (now underway in Ireland) where the law was recently passed. One might question, is the legal abortion the result of 9/11? Or what are the lessons the classic mythology offers?

Countless mythical narratives, ranging from Christ to Gilgamesh, center on the story of heroes who not only are in a difficult position but they ran the risk of being killed by their infancies. They come from a royal line that disputes the power of the established monarchy. Like Mary, the ladies who bring these heroes to life are compelled and forced unilaterally by the divinity. Although Mary does not want the situation and risks she will go through, she accepts God's decision. What is more important, she offers her hospitality to the new child in the same way the host gives hospitality to the unknown guests. This, doubtless, seems to be the lesson mythology gives. The mother works as a symbolic host feeding and protecting her offspring as long as a specific period of time. In consonance with Jacques Derrida, hospitality takes two contrasting forms. While the restricted hospitality signals to giving-while-receiving ritual to those who can pay for that, the unconditioned hospitality is given without asking anything in return (Derrida & Dufourmantelle, 2000). The fear inspired by terrorism not only closes the nations but also affects the hospitality in the Western civilization (Korstanje, 2017). The modern society, where the undesired "Other" has no room, does not work pretty different from the mother who opts for ending her pregnancy. She is afraid because the future child confronts her own comfort. In the same way, terrorism echoes an extortive logic which produces an extreme sentiment of panic and aversion for everything that cannot be disciplined and controlled. The society is reluctant to welcome the stranger who needs protection and asylum because it interrogates the established sense of normalcy. As Daniel Innerarity (2017) puts it, risk and the figure of the foreigner are inextricably intertwined. While risks loom and appear suddenly, the foreign traveler asks for the law of hospitality transforming our sense of security. To some extent, West is between the wall and the deep blue sea. The obsession for the total control (zero-risk society) despoils humans from their essence: the contingency. In a nutshell, the fear of terrorism is liberating a radical cosmology aimed at neglecting the "Other" and one of the cultural touchstone of Western civilization: the hospitality.

Conclusion

The mother hosts her child in the same way the society grants hospitality to strangers. The abortion symbolizes the lack of tolerance for uncertainness and the contingency. The new unborn is fully rejected in the same way a Syrian refugee is jailed and deported in Europe. Whether the terrorist sacrifices himself to kill others, the modern mother assassinates the unknown "Other" to protect herself. The terrorist melts the individuality of his being to create mass-death, while the other subordinates the collectivity into her own individualist wish: not to be a mother. The Western woman does not accept the interpellations of the destiny, and she wants to dispose of her own body. As discussed, terrorism accelerated the times affecting not only the conception about strangers but also the social bondage. This chapter explored the dilemma of abortion (without inferring in any moral judgment) as the mark of the decline of hospitality. We hold the thesis that the modern self has serious problems to understand the alterity when it confronts the own ontological security. The right of legal abortion should be framed as the rise of a new climate of anti-hospitality imposed by modern terrorism and the culture of fear.

REFERENCES

Abbas, T. (2012). The Symbiotic Relationship Between Islamophobia and Radicalisation. *Critical Studies on Terrorism*, 5(3), 345–358.
Altheide, D. (2017). *Terrorism and the Politics of Fear*. New York: Rowman & Littlefield.
Andersen, M. L., & Taylor, H. F. (2007). *Sociology: Understanding a Diverse Society, Updated*. London: Cengage Learning.
Cheyfitz, E. (1993). Savage Law, the Plot Against American Indians in Johnson & Graham's Lesee v MÍntosh and the Pioneers. In A. Kaplan & D. Pease (Eds.), *Cultures of United States Imperialism* (pp. 109–128). Durham: Duke University Press.
Chomsky, N. (2015). *Pirates and Emperors, Old and New: International Terrorism in the Real World*. New York: Haymarket Books.
Clark, W. R. (2005). *Petrodollar Warfare: Oil, Iraq and the Future of the Dollar* (p. 31). Gabriola Island: New Society Publishers.
Coleman, S. (2013). Actors of History? Religion, Politics, and Reality Within the Protestant Right in America. In G. Lindquist & D. Handelman (Eds.), *Religion, Politics & Globalization: Anthropological Approaches* (pp. 171–188). Oxford: Berghahn.
Cott, N. F. (1987). *The Grounding of Modern Feminism*. New Haven: Yale University Press.
Derrida, J., & Dufourmantelle, A. (2000). *Of Hospitality: Cultural Memory in the Present* (R. Bowlby, Trans.). Stanford: Stanford University Press.
Dooling, A. D. (2005). *Women's Literary Feminism in 20th-Century China*. London: Macmillan.
Doumani, B. (2006). Between Coercion and Privatization: Academic Freedom in the Twenty-First Century. In B. Doumani (Ed.), *Academic Freedom After September 11* (pp. 11–60). New York: Zone Books.
Fassin, D. (2018). *Por una Repolitizacion del mundo: las vidas descartables como desafio del siglo XXI (Through a New Politics in the World: Wasting Lives as a Challenge for XXIth Century)*. Buenos Aires: Siglo XXI.
Feierstein, D. (2017). *Genocide as Social Practice: Reorganizing Society Under the Nazis and Argentina's Military Juntas*. New Brunswick: Rutgers University Press.
FitzGerald, F. (1986). *Cities on a Hill: A Journey Through Contemporary American Cultures*. New York: Simon and Schuster.

Freedman, E. B. (2003). *No Turning Back: The History of Feminism and the Future of Women*. London: Ballantine Books.

Giddens, A. (2013). *The Transformation of Intimacy: Sexuality, Love and Eroticism in Modern Societies*. New York: John Wiley & Sons.

Gronke, P., Rejali, D., Drenguis, D., Hicks, J., Miller, P., & Nakayama, B. (2010). US Public Opinion on Torture, 2001–2009. *PS: Political Science & Politics*, 43(3), 437–444.

Guidotti-Hernández, N. M. (2011). *Unspeakable Violence: Remapping US and Mexican National Imaginaries*. Durham: Duke University Press.

Harding, S. G. (1986). *The Science Question in Feminism*. Ithaca: Cornell University Press.

Harris, M. (2001). *The Rise of Anthropological Theory: A History of Theories of Culture*. Walnut Creek: AltaMira Press.

Hawkesworth, M. (2006). *Globalization and Feminist Activism*. New York: Rowman & Littlefield.

Hobbes, T. (1968). *Leviathan: Edited with an Introduction by CB Macpherson*. New York: Penguin Books.

Hofstadter, R. (1992). *Social Darwinism in American Thought*. Boston: Beacon Press.

Howie, L. (2007). The Terrorism Threat and Managing Workplaces. *Disaster Prevention and Management: An International Journal*, 16(1), 70–78.

Howie, L. (2017). *Terrorism, the Worker and the City: Simulations and Security in a Time of Terror*. Abingdon: Routledge.

Ignatieff, M. (Ed.). (2009). *American Exceptionalism and Human Rights*. Princeton: Princeton University Press.

Ignatieff, M. (2013). *The Lesser Evil: Political Ethics in an Age of Terror*. Princeton: Princeton University Press.

Ignatiev, N. (1995). *How the Irish Became White*. New York: Routledge.

Innerarity, D. (2017). *Ethics of Hospitality*. Abingdon: Routledge.

Janiewski, D. (1991). Southern Honor, Southern Dishonor: Managerial Ideology and the Construction of Gender, Race and Class Relations in Southern Industry. In A. Baron (Ed.), *Work Engendered. Toward a New History of American Labor* (pp. 47–69). Ithaca: Cornell University Press.

Korstanje, M. (2013). Empire and Democracy, A Critical Reading of Michael Ignatieff. *Nómadas: revista crítica de Ciencias Sociales y Jurídicas*, 38(2), 69–78.

Korstanje, M. (2016a). *The Rise of Thana Capitalism and Tourism*. Abingdon: Routledge.

Korstanje, M. E. (2016b). *Terrorism in a Global Village*. New York: Nova Science Publishers.

Korstanje, M. E. (2017). *Terrorism, Tourism and the End of Hospitality in the 'West'*. New York: Springer.

Korstanje, M. (2019). *The Challenges of Democracy in the War on Terror: The Liberal State Before Terrorism*. Abingdon: Routledge.
Lasch, C. (1979). *The Culture of Narcissism*. New York: Warner Books.
Leslie, H., & Drake, J. (1997). *Third Wave Agenda: Being Feminist, Doing Feminism*. Minneapolis: University of Minnesota Press.
Lipset, S. M. (1997). *American Exceptionalism: A Double-Edged Sword*. New York: WW Norton & Company.
Lyon, D., & Bauman, Z. (2013). *Liquid Surveillance: A Conversation*. New York: John Wiley & Sons.
Madsen, D. L. (1998). *American Exceptionalism*. Jackson: University Press of Mississippi.
Messer-Davidow, E. (2002). *Disciplining Feminism: From Social Activism to Academic Discourse*. Durham: Duke University Press.
Miller, P. (1953). *The New England Mind from Colony to Province*. Cambridge: Harvard University Press.
O'Brien, E. (2004). Torture and the War on Terror. *Social Education*, 68(7), 453–456.
Pease, D. E. (1987). *Visionary Compacts: American Renaissance Writings in Cultural Context*. Madison: University of Wisconsin Press.
Poole, E. (2002). *Reporting Islam: Media Representations and British Muslims*. London: IB Tauris.
Post, R. (2006). The Structure of Academic Freedom. In B. Doumani (Ed.), *Academic Freedom After September 11* (pp. 61–106). New York: Zone Books.
Puar, J. K. (2017). *Terrorist Assemblages: Homonationalism in Queer Times*. Durham: Duke University Press.
Robin, C. (2004). *Fear: The History of a Political Idea*. Oxford: Oxford University Press.
Roth, K. (2006). Torture in the War on Terror. *Harvard International Review*, 28(2), 80.
Saaed, A. (2005). *Identity Papers: A Journal of British and Irish Studies*, 1(1), 15–31.
Skoll, G. R. (2016). *Globalization of American Fear Culture: The Empire in the Twenty-First Century*. New York: Palgrave Macmillan.
Skoll, G. R., & Korstanje, M. E. (2013). Constructing an American Fear Culture from Red Scares to Terrorism. *International Journal of Human Rights and Constitutional Studies*, 1(4), 341–364.
Stilz, A. (2009). *Liberal Loyalty: Freedom, Obligation & the State*. Princeton: Princeton University Press.

Timmermann, F. (2014). *El gran terror. Miedo, emoción y discurso. Chile, 1973–1980 (The Great Terror: Fear, Emotions and Discourse, Chile 1973–1980)*. Santiago de Chile: Copygraph.
Tyrrell, I. (1991). American Exceptionalism in an Age of International History. *The American Historical Review, 96*, 1031–1055.
Weber, M. (1958). *Essays in Sociology*. New York: Oxford University Press.
Weber, M. (2013). *The Protestant Ethic and the Spirit of Capitalism*. Abingdon: Routledge.
Žižek, S. (2014). *The Universal Exception*. London: Bloomsbury Publishing.

CHAPTER 4

The War on Terror

INTRODUCTION

On September 11, 2001, the world changed forever. Four civilian airplanes were weaponized against the World Trade Center and Pentagon, and a fourth domestic flight (UA 93) was forcefully grounded in Pennsylvania. All passengers and aircrew members lost their lives. As a result of these combined blows, 2966 innocent civilians lost their lives, while others 6000 were injured. This tragedy was known as 9/11 and of course started a new era where terror and anxiety prevailed (Aly & Green, 2010; Holloway, 2008; Howie, 2012; Skoll, 2016). Days later to this terrorist attack, which was perpetrated by Osama Bin Laden and his group Al-Qaeda, George W. Bush (former president of the US) declared the "War against Terrorism", which was known as "the War on Terror". Although in May of 2013, Barak Obama pointed out that the War on Terror was over, some specialists agree that its effects continue to date (Kellner, 2003; Korstanje, 2017, 2018). The present chapter discusses critically the allegories formed behind 9/11 and "the War on Terror", as well as the limitations of preventive war. This work synthesizes almost one decade of debate, reflection and insights revolving around terrorism and 9/11. Most likely, one of the aspects that moved us to write this chapter was the solidarity expressed by other nations to the US after this attack. To set an example, Argentina faced two serious terrorist attacks in Buenos Aires during 1992 and 1994, like many other cases in the world. Today's

Argentinians feel that terrorism is a problem limited to the US and Europe. Mysteriously, others terrorist attacks occurred in the periphery were symbolically subordinated to 9/11 as a founding event that claimed the rights of the American government to intervene in the politics of other autonomous countries. Gilbert Achcar (2015) calls the attention about a type of "narcissist commiseration", which suggests that the feeling of solidarity expressed in favor of the US exhibits an ethnocentric discourse, where the peripheral nations internalize the sentiments of the center to be part of the "chosen peoples". Quite aside from this, we hold the thesis that terrorism operates within the discursivity of language, which is historically enrooted in the American past. Terrorism functions as a "phantom" that looming from the past interrogates the repressed guilt in the present time. To put this in other terms, the fear in Americans to be victims of mass destruction weapons situates as an external object internally associated to a much deeper repression by the two atomic bombs that destroyed Nagasaki and Hiroshima. In psychological terms, *the projection* can be defined as a "mechanism of defense" where the undesired feeling, which is repressed by the self, is projected against an external object (or threat). A common form of projection at the time a subject, overwhelmed by the innermost angry feelings, blames others for potential hostility. The same applies "to the phantom of terrorism", which the narratives of 9/11 alimented.

Sociology After 9/11

Although many scholars believe that 9/11 was a "founding event", which means a significant event that in some way changed the geopolitics (Entman, 2003; Pyszczynski, Solomon, & Greenberg, 2003; Weiss, 2006), other scholars hold that it represented a "simulacrum" articulated to pose polices otherwise would be rejected (Baudrillard, 2003; Klein, 2007; Skoll, 2016). In his book *The Spirit of Terrorism*, French philosopher Jean Baudrillard (2003) acknowledges the importance of discussing terrorism as encapsulated in the world of media and digital simulations. He dubbed 9/11 as "the mother event", a pure event which subordinates all previous similar events paving the ways for the rise of hyper-reality and "pseudo-events". Following Baudrillard, the world as we know sets the pace to a new stage, where historical events are symbolically transformed in pseudo-events. To understand this, Baudrillard cites the plot of Spielberg's film *Minority Report* where a futurist society is in a frenetic quest for eliminating the local crime. The precogs, who previsualize the future or the crime before it is committed, help police to arrest potential criminals. As a result of this, the rate of crimes slumped down to zero, while would-be murderers are imprisoned in a climate of virtual reality. In fact, this not only defies the essence of Roman jurisprudence but also introduces future to rule the present time. Baudrillard eloquently confirms that the fear of terrorism leads very well to a renounce to live in the present, where events (risks) are imagined and mitigated earlier than they really take place in the present. This is the moment when a "historical event" mutates toward a "pseudo-event" (Baudrillard, 2003). Some other voices toyed with the belief that terrorism opened the doors to a "virtual world" which is stimulated by the digital technology. In this vein, P. Virilio (2010) argues convincingly that the global capitalism is far from solving the created risks introducing the available technology and scientific thinking. Rather, scientists devote their efforts and time to forecast those global risks which may place the capitalist system (if not the market) in jeopardy. Instead of promoting the collective well-being, scientists are moved to protect the interests of the capital market. In this respect, M. Augé (2002) contends that the media has created a climate of anxiety and mistrust that obscures the causality of terrorism. The professional journalism disposed of terrorism and the War on Terror as an ideological platform (spectacles), where images continuously circulate replacing the previous one. In this way, the audience is hand-tied to the order of events, losing their understanding of what they are gazing. Slavoj Žižek, who does not need presentation, publishes in 2015 a book which merits our attention, *The Universal Exception*. In this work, he reminds that the doctrine of the economic scarcity is the symbolic touchstone of capitalism. This suggests that the system encourages poverty and frustration as a form of efficient governance. Starting from the premise that the digital world serves as an ideological cage where the frustrations of citizens are revitalized and sanitized, Žižek reminds particularly how 9/11 and terrorism woke the Western civilization from the slumber it was in. In psychoanalytical terms, this event was the irruption of the reality principle into the wonderland of consumption (Žižek, 2015).

Last but not least, we cannot close the section without discussing the intersection of terror and "the preventive attack". As G. Skoll brilliantly wrote, the American imperialism has developed a derogatory view of the non-Western "Other" cultivating anxiety and fear as two guiding forces in politics. For all those who wanted to escape to the logic of consumption, the introduction of fear was a dissuasive instrument. The legitimacy of the ruling elite centered on fear as a disciplinary mechanism of control. In view of this, the globalization of capitalism liberated this constrained fear beyond the borders of the US. This is exactly what represented 9/11 and the "War on Terror". Skoll understands that the archetype of terrorism allows the imposition of neoliberal policies aimed at disarticulating the worker unions and in consequence vulnerating the work-force worldwide (Skoll, 2016). To some extent, terrorism and the labor precaritization are inevitably entwined.

THE PREVENTIVE WARS

To date, there is a great controversy revolving around who is the real perpetrator of 9/11. A seminal book in this direction was originally authored by David. R Griffin under the title, *The New Pearl Harbor*. In this text, the author dangles the possibility that the attack was allowed or planned by the US government (Griffin, 2004). Quite aside from this, what specialists agree seems to be that this "founding event" legitimized the two US-led invasions to Afghanistan and Iraq, accelerating what would be a historical tragedy in international politics (Kagan, 2004; Weisberg, 2008). The US was not only widely criticized for the interventions in other sovereign nations—like Iraq—but also confronted with the idea of a "sustainable security" (Suri & Valentino, 2016). While the US launched to monitor what specialists dub as "rogue states" just after the end of WWII, it is no less true that the stock and market crisis that happened in 2008 limited the resources of many capitalist economies, forcing authorities to rethink their security-related strategies. In 2004, Kagan alerted on the risks of the US acting as a watch-dog of the global order, when Europe was cultivating peace and trade as proper forms of relations. In this respect, William Inboden (2016) suggests that the model of national security consists in moving the resources to protect the borderlands and borders checkpoints to defend the citizens. However, what does happen when the same resources are unilaterally targeted to protect others nations? This is doubtless a good question that leads Inboden to accept that:

> Two paradoxes stand out about George H. W Bush administration' national security policy and institutions. First, while the Bush administration is known for a substantially different foreign policy approach than the Reagan administration, it largely maintained the same national security institutions that were inherited from the Reagan White House. Second, while the Bush

administration presided for four years over some of the most profound shifts in the international order that the world history had ever seen, the administration largely eschewed any profound institutional reform within its own government. (Inboden, 2016: 152)

The above-cited excerpt reveals two important aspects of the role of the US as the police coming across the world. On the one hand, Americans historically forged a spirit of suspicion or mistrust regarding the [native] "Other", which gradually ushered them into what Kelman names as "counterfeit politics" (Kelman, 2012). The narratives of conspiracy are constructed to see the external world as a dangerous place to live. Democracy and the US have the "divine mandate" to protect the emerging nations from the claws of dictators. Particularly, this is a problematic question simply because "the right of self-representation" which accompanies any sovereign state is violated by the US in the name of democracy, while in other cases, the US reserves the act of being interrogated by another state (invoking the same right of self-representation) (Korstanje, 2018). What is more important, as David Altheide brilliantly observed, such a climate of paranoia, which was originally instilled by the Puritans, paved the ways for the articulation of "preventive attacks or the doctrine of preemption" as a valid form of self-defense. Altheide laments that the obsessions for security and safety are day to day shifting the democratic institutions, accepting unethical practices such a torture or the habeas corpus suspensions, otherwise would be legally rejected. The precautionary principle, from where the needs of preventive wars are legitimized, rests on a political vacuum that confuses the legal analysts. To what extent is the declaration of a war against a second state constitutional without firm bases? Preemption suggests that any state has the right to declare war against another state on the basis of immediacy of a potential attack. However, as Altheide probed, this principle can be politically manipulated for the ruling elite to prevent the social change, or internally to placate the social discontent (Altheide, 2017). On the other hand, these needs of expansion and control have unilaterally caused a serious economic crisis that today forces the analysts to revisit entirely the National Security program (Stiglitz, 2008). Corey Robin (2004) traces back an interesting review on the history of fear in the West. With the benefits of hindsight, he finds that the fear plays a paralyzing role not only silencing political opposition but also putting lay-citizens asunder from the political arena. As a result of this, the function of fear is double-fold. While fear silences the internal discontent, it is equally true that fear projects in the external enemy an object to achieve the social cohesion. Lastly, P. Gray (2007) exerts a radical criticism of the limitations of preventive wars. He toys with the belief that the notion of *preemption* was historically distorted according to many interpretations and interests. The problem is not given to what extent "preemption" is efficient to prevent any external attack, but in fact, the point is that to what extent analysts understand the difference between *prevention and preemption*. Surely, the sudden strike that knocked the US (just after 9/11) prompted a climate of anxiety as never before. Since the worst has happened, there would be a question of time America will be struck by terrorism again, the social imaginary precluded. As Gray clarifies, *preemption* signals to any potential attack which is underway, or gains further credibility to be imminent. As long as the Cold War, the idea of an imminent attack was on the agenda of governments, but in fact, preemption is enrooted to an abstract concept, which is very hard to precise. After all, the looming strike would never take room. Rather, prevention refers to a major strategic concept oriented to deter a condition of affairs in which case it is too late to protect itself. To put this in bluntly, any passive action, which means the decision made by the Executive Branch not to attack other hostile states, entails a taken-for-granted invasion. The essential difference between preemption and prevention lies in the fact that the former situates as a political and ineludible option when the war is certain, while the latter appeals to the preventer's choice, which can attack first or tolerate an external attack to proceed.

Alex Bellamy (2006) states the right of just wars to legalize when the force should be used. This is not a new point; it was debated by ancient philosophers and thinkers from the immemorial times. The just war gives theorists a bunch of criteria to rationalize the use of violence. These notions and articulation, far from being static, vary on time and culture. Bellamy explores the evolution of just war tradition from the "holy-war" toward the creation of law. In sum, there are no clear-cut borders between the definitions of preemption and prevention, but what is clear seems to be that the "War on Terror", as well as all allegories around it, says little with respect to the quest of an absolute secure society.

The Problem of the Precautionary Principle

Seven days later the attack to New York, G. W. Bush declared the war against terrorism, which was known as "the War on Terror". Some scholars—at the turn of the twenty-first century—alluded to the risk-management theory combining different sources of information to locate and eradicate those risks (like terrorism) that threaten society (Amoore & De Goede, 2008; Aradau & Van Munster, 2007; Ericson & Doyle, 2004). Although terrorism was a major threat for the US even earlier than 2001, it was not until the attack on the World Trade Center that the theme captivated not only specialists, publishing thousands of books in the year, but the "global audiences" (Altheide, 2017; Eid, 2014; Korstanje, 2018). Terrorism was anglicized at the same time the US reserved the right to be a leading nation that explains to the world what terrorism is (Korstanje, 2017).

As the previous argument given, Andrew Hoskins and Ben O'Loughlin (2009) speak of the "cultures of immediacy" to denote those global audiences which gaze "terror" as a form of entertainment. Despite the distances, terrorism-related news is disseminated and circulated across the globe in seconds. The CNN coverage on the War on Terror continues the dominant stereotypes and prejudices which are orchestrated according to a *modulation of terror*. The collapse of Soviet Union, as well as the rise of new emergent crises, puts the US against one of its hardest dilemma: *how can the future be efficiently controlled?*

> We propose two concepts that give us analytical leverage to understand this crisis. The first is the modulation of terror. News modulates terror by often simultaneously amplifying and containing representations of threat. News amplifies by inflating the seriousness of threats by connecting a single threat to others, or by representing threat in vague, indefinite terms through speculation…. (p. 14)

In this token, Professor Jared Ahmad contends that though terrorism is a real threat for the West, the media has elaborated a specific discursivity around it. Just after the attacks on World Trade Center and Pentagon, the Press introduced new narratives (allegories) aimed at stressing the figure of Osama Bin Laden and his radical group, Al-Qaeda. He focuses his study on the coverage of BBC from 2001 to 2011, the exact year when Bin Laden is killed by US Special Forces. Ahmed discusses critically to what extent Al-Qaeda becomes a representational product, in which case its degree of danger can be exclusively understood within the "discursive entity". The media coverage, though it claims for the sense of objectivity, follows the previous stereotypes and prejudices revolving around "Orientalism", and "Middle East", from the colonial rule. Arabs and Muslims are often portrayed as "enemies of democracy", or "authoritarians" whose hearts and minds are incompatible with the individual rights. In fact, as Ahmad said, the War on Terror should be interpreted as an encounter of signs, discourses and political ideological resources that limits the public discussion to a binomial logic, *us* versus *them*.

W. Soyinka (2005) reminds how the citizens of the Third World were not surprised when Al-Qaeda humiliated the US. The climate of fear and violence is part of the landscape of African nations. In this respect, he claims that one of the aspects of global power that facilitates the feeling of uncertainty seems to be the lack of a visible rivalry once the USSR collapsed. The politic terror promulgated by states diminishes the dignity of enemies. These practices are rooted inside a territory but paved the way for a new form of terrorism which ended in the World Trade Center attacks. It is incorrect to see 9/11 as the beginning of a new fear but as the latest demonstration of the power of an empire over the rest of the world. Mass communications, though, transformed our ways of perceiving terrorism even if it did not alter the conditions that facilitate the new state of war. At a closer look, Soyinka examines the current connection between power and freedom. Unlike classical totalitarian States which are constructed by means of material asymmetries, the quasi-States construct their legitimacy by denouncing the injustices of the World. Quasi-States are not only terrorist cells but also mega-corporations which work in complicity producing weapons for one side or the other. Making profit of human suffering is a primary aspect that characterizes these quasi-states. The uncertainty these corporations engender denies the minimum codes of war by emphasizing the inexistence of boundaries and responsibilities. Once rectitude has been substituted by the right to exercise power, pathways toward a moral superiority are frustrated. Unlike the disaster of the Napalm-bombing of non-combatants by the US in Vietnam, this new War on Terror is characterized by targeting innocents as a primary option. In opposition to conventional wars, War on Terror expands fear under the following two assumptions: (a) hits can take place anywhere and anytime, and (b) there is no limit to brutality against non-combatants. Wars depend on the capacity to control others based on the principle of power. Governments often need the material resources of their neighbors. Where the expropriation method of capitalist trade fails, war finds success. One might speculate that war should be understood as an extension of economic production. The role played by fear in late modernity is rooted in a desire for domination that has nothing to do with religiosity or even to religious fundamentalism, which in recent years has become synonymous with cruelty (Soyinka, 2005).

In other approaches, we said that the liberal state started a war against terror, but it is an emotion which takes an abstract nature. The wars are against specific objects or enemies. Unlike other conventional wars, terrorism opens the doors to vague risks, which are fulfilled according to the dynamics of global capitalism. It is remarkable that terrorism has been commoditized as an entertainment industry that results in video games, movies, documentaries and so forth, but paradoxically less is known on its nature. The War on Terror denotes the possibility we are struggling against an emotion, *the fear*. This suggests that any war against fear is futile because it is part of our nature. Still further, terrorism gradually undermines one of the symbolic touchstones of Western civilization: the sacred law of hospitality (Korstanje, 2018, 2019). This begs some interesting questions, what is the War on Terror? And in what way does it change the daily life in Occident?

How Terrorism Is Changing Us

Jean Baudrillard calls the attention on the significance of 9/11 for the social imaginaries. In a world of hyper-consumption, where the non-Western "Others" is humiliated, terrorism should be considered as something else than a simple attack or a declaration of war. Terrorism is a humiliation inflicted over the most powerful nation in the World (Baudrillard, 2003). Douglas Kellner (2005) clarifies that terrorism is like a game whose rules are completely broken. One of the most important contributions of Baudrillard to the study of terrorism was undoubtedly associated to decipher the meaning of 9/11 and the subsequent War on Terror. The globalization triumphed over others forms of classic wars that castigated the mankind in the past, and of course, this founding event unfolded the "fourth world war". The globalization is in Baudrillard's eyes the opposite to a force that promotes democracy and material prosperity. Rather it generates a hyper-standardization, commoditization and homogenization of human action.

As the previous backdrop, Korstanje wrote that it is very hard to imagine terrorism beyond the hyper-mobile world we inhabit today. In fact, globalization and terrorism are inextricably intertwined. To put this slightly in other terms, terrorism consternated the social imaginary of the West because it employed the mass-transport which was the pride of capitalism as real weapons placing the most important nation in this world in jeopardy (Korstanje, 2019). The archetype of the World Trade Center encourages awareness about mobilities and tourism, which were two key factors in the configuration of the modern nation-state, and which were weaponized against civilian and military-targets.

In the Roman Empire, the legal figure of *dictator (magister populi)* inscribes in the political needs of suspending temporarily the law, and the division of powers in favor of a leading ruler who is entitled to face the external threat. Over years, many philosophers have theorized on the interplay between security and republicanism. The efforts to secure the territory are sometimes unconducive to the correct separation of branches. This seems to be exactly the point of departure of John Owens and Riccardo Pelizzo in his paper *The Impact of the War on Terror on Executive-Legislative Relations*. One of the scholars who publicly defended the concentration of power in few hands in moments of crises was C. Schmitt. From his viewpoint, no form of government survives when a danger is imminent. Owens and Pelizzo acknowledge that the attacks in New York, which resulted in the declaration of war against terrorism, imply a global state of emergency similarly to what Schmitt and other analysts mention. As authors go on to admit:

> Just over a month after 9/11, an infamous, now rescinded, secret legal opinion from the Bush administration's Office of Legal Counsel brazenly asserted that in order to preserve national security and pursue the so-called 'war on terror' 'the government's compelling interests in wartime' justified the curtailing of civil liberties in the US, including free speech, press rights, and privacy and search and seizure protections. (Owens & Pelizzo, 2010: 123)

Most likely, the US subordinated to mobilize other nations "to the fight against terrorism" as a unilateral condition of friendship. You are with us or against us! G. W. Bush declared. This created an international alliance that legitimated the intervention in Afghanistan. Years later and just after the failure to dismantle the MDW (mass destruction weapons) Saddam Hussein supposedly possessed, the US is brought into disrepute. This point reminds that beyond the urgency and the needs of reformulating the constitutional rights of citizens, there is a twilight zone that marks the borders between dictatorship and democracy. The main impact on this war was certainly located in the legal jurisprudence and the subordination of Parliament to pass laws that vulnerated the rights of those suspected of terrorism. Another important aspect that explains how terrorism changes us is the emergence and expansion of chauvinist expression or even patriotism which act as a mechanism to control the anxieties lay-people usually feel. This happens because before a critical event politicians and citizens overemphasize the power of causal attributions. As Gail Sahar explains, the obsession for grasping the situation leads toward prediction, or projections which are intended to blame others for what they experience. The theory of attribution says that:

> Individuals frequently search for causes of events that happen to themselves and others, particularly surprising or negative events (e.g., see Weiner, 1995). Understanding the cause of a negative event helps one to place blame on the responsible parties and respond appropriately. Judgments of causal responsibility are informed by ideological orientation. (Sahar, 2008: 198)

The role of patriotism in these processes is of paramount importance in encouraging or discouraging hostility against an ethnic minority. The results obtained by Sahar in a sample of students reveal three important points. Americans elaborate attributions (discourses) that explain the virulent violence of 9/11, forming three clear-cut axes: (a) the inefficient American foreign policy, (b) resentment against the US's economic prosperity and (c) proper psychological frustrations or pathologies in the terrorists' minds. Though these narratives change over time, Sahar alerts about the risks of blaming others for the external events. As a collective, Americans not only recognized themselves as part of the same group, but the created *attribution* protected the US of any charge—even the misleading positions of other administrations in international affairs.

Gabriel Weimann (2005) holds that terrorism follows the patterns of media attention. In spite of the efforts of the government to prohibit some radicalized cells from accessing mass media, it is equally true that the emerging media technology is too diffuse and broad and can be used by terrorists without any limits. More easily transmitted, media plays a leading role in disseminating the terrorist's message to a wider network of users and viewers. This is one of the challenges the counter-terrorism policies should keep in mind in the years to come. Terrorists look different tactics to affect the credibility of the state. On a closer look, the "last option" tactic (or no choice motive following Weimann) signals to locate violence as the only feasible way to defeat an always oppressive enemy. The second tactic refers to *a de-legitimization of the force* in view of the fact that terrorists are portrayed as fighters who against their will are obliged to confront the state. Third, terrorists normally exaggerate a situation of vulnerability justifying violence as a "the struggle of the weaker". The fear terrorists instill in society is repressed by the security forces. In sum, all these (radicalized) narratives may escalate to an uncontrolled state of violence, in which case the democratic institutions set the pace to a more repressive state. As Weimann puts it:

> The emergence of media-oriented terrorism presents a tough challenge to democratic societies and their liberal values. The threat is not limited to media manipulation and psychological warfare launched by terrorists; it also includes the danger of restrictions imposed on the freedom of the press and freedom of expression by those who try to fight terrorism. (Weimann, 2005: 389)

The above-cited paragraph leads F. Sheth (2011) to adopt the Foucauldian legacy to understand how after 9/11, the political regime manipulated emotions as love, fear and anxiety to impose a disciplinary mechanism to change how the power is administered. Finally, we cannot close this section without discussing the role of Marxism and the conspiracy theory in this much deep-seated issue. Unlike other studies, Marxists emphasize on the material consequences of terrorism. When we mean "material consequences", we say the economic asymmetries which are imposed by capital-owners to exploit the workers. Terrorism permits the emergence and circulation of ideological narratives that hide the causes of labor exploitation. Hence, these theorists start from the premise that workers do not conspire, but only receive passively the conspiracy theory as a reality. In view of this problem, David MacGregor and Paul Zarembka (2010) debate critically the current position of Marxists with respect to the worker's participation in politics. The Marxist theory indicates that the state works following the guidelines of the dominant class. There are some limitations that place Marxists into a conceptual gridlock. One of them, to cite anyone, aims to stress 9/11 as a planned attack that happened officially validated by the US government. They introduce conspiracy—as an emptied signifier—which is molded according to the individual projections. But as Marx has amply recognized, the conspiracy aligns with the divide and rule tactic. Conspiracy not only corresponds with a covered side of ideology, but also impedes the workers to grasp reality as it really is. However, as the authors noted, Marx adopted an ambiguous position. While he systematically compared conspiracy to ideology, he held that capital-owners need to cover the state of exploitation through the articulation of (false) bourgeois representations. This poses the suspicion in the direction of the ruling class.

> Drawing both on Marx's own writings and on the notion of deep politics, we have argued that conspiracy by members of ruling elites is intrinsic to capitalism. Regarding 9–11, we urge an articulated distinction within Marxism between acts of the state that are public and open to direct investigation, and acts of the state that are concealed, secret, and indeed conspiratorial. We do not believe that this distinction has been previously incorporated into Marxist theory. Our paper then offers a view of 9–11, inspired by this perspective. That is, while 9–11 was clearly the result of a conspiracy, we put forward the consideration of a domestic conspiracy, still being hidden, at the expense of the exclusively foreign, bin Laden conspiracy that has been claimed officially. (MacGregor & Zarembka, 2010: 158)

In his book *Counterfeit Politics, Secret Plots and Conspiracy Narratives in Americas*, David Kelman realizes that far from being a problem of left-wing writers, the conspiracy narratives are enrooted in the modern politics. Over years, scholars thought the secret plots as a pathological mode of politics. Centered on the literary works of Piglia and Borges, he shows how the political machinations are part of the politics and the distribution of power. The conspiracy theory involves the US and Latin America crossing the borders of cultures. Conspiracy consists in the legitimization of silence, which through the secrecy recreates two alternative circuits: one official and the other unofficial. While the former says something about the world, the latter signals to a conspirational fable (an empty signifier) which is manipulated to achieve the in-group cohesion. Equally important, across all modern democracies, politics appears to embrace the conspiracy plot to validate the official unilateral discourse dictated by authorities.

Basically, Kelman's view rests on two troublesome points. Firstly, the figure of conspiracy closes the doors for the questions placed once the ideology is directly confronted. The theory of conspiracy is articulated to keep the interests of the status quo no matter the nature of facts. We hold the thesis that conspiracy and ideology are two sides of the same coin. Let's cite readers the example of 9/11 and terrorism. The ideology fabricated and transmitted to the world says that the US is a mega-political power, whose hegemony, after the Soviet bloc collapse, reaches the entire world. This American power, from its inception, attempted to expand egalitarian values and democracy to all countries to strengthen the well-being of their citizens. But if the US is this benevolent all-seeing eye, why was 9/11 not prevented? A tentative, conspiratorial, answer would be that American officials were familiar with the possibilities of an attack and chose to allow it to happen. Here, the conspiracy theory represents the other pole of ideology. It fills in for the missing piece of the ideological position. Any conspiracy exhibits valid efforts to make controllable what is in nature uncontrollable. Whenever reality overrides the fiction, the theory of conspiracy responds to the questions that ideology keeps open. Accusations of conspiracy theories work as a mechanism to validate the official discourse, silencing those voices which may bring a radical shift to the system. Conspiracy theory functions as an epithet which plays an important part in ideology, not social or political theory. Last but not least, in psychology, conspiracy and paranoia are not pathologies; rather, they are adaptive mechanisms of the frustrated ego to an intransigent environment. Conspiracy theories surface to explain the failures of the ego to reach life goals, and they can lead to the rejection of the "reality principle".

Conclusion

In this chapter, we outlined the important points that left the archetype of 9/11 and terrorism. Through different sections, we avoided any moral position or judgments which obscure more than they clarify. As Luke Howie (2012) puts it, we are not terrorists while we are not taking part in terrorist cells; as scientists, we can claim to know only about how terrorism is changing our society. The society condemns terrorism considering their acts as a product of hatred-filled hearts. For ethnographers, like us, this represents a serious limitation because we cannot be in contact with terrorist groups. Any vital information or dialogue would be asked for the police or the security forces. Hence, we can discuss—as objectively as possible—the effects of terrorism in daily life. In view of this, the present chapter explored theoretically the narratives and allegories revolving around the archetype of 9/11 and the failed "War on Terror". The first point of entry in this discussion seems to be the presence of fear, and extortion which instrumentalizes the other's pain to achieve the proper (corporative) goals. As Howie cited, terrorists do not want a lot of people dying, they want rather a lot of people watching. To some extent, the terrorist attack intersects with the needs of notoriety and publicity only the (mass) media offers. We have interrogated on the connection of terrorists—who often move like celebrities—and the oxygen given by the media.

To the best of our knowledge, it exhibits a fertile ground for terrorism to air their claims while the media gains further investors. For some reasons which are very hard to elaborate upon here, our society is captivated by what Baudrillard dubbed as "the Spectacle of Disaster", or Korstanje named as "Thana-capitalism", a new stage of capitalism where the Other's death is the main commodity. Third, the act of terrorism—besides being a form of entertainment—was a humiliation to the hegemony of West. The fear instilled days after 9/11 was associated with the fact that the most powerful nation was suddenly attacked taking a tactic advantage of the mass transport system. Tourism and mobilities which to date were motives of pride for the US were weaponized against vulnerable targets—if not innocent workers. The fourth element of this shocking event derived from the needs of introducing conspiracy narratives to keep the machine working. As discussed in the preceding section, the 9/11 was much what the world can digest, and the belief the US government planned the attacks restored the harmony to the system. While the US was imagined as the strongest economic and military force of the world, ideology places Americans as a part of the "chosen people" (i.e., what at the bottom the theory of exemption proclaims: we, Americans are different, special!). However, sooner or later the reality principle overrides our dreams. This evinces that the power of ideology exhausts in the same way the legitimacy of the ruling elite weakens. To avoid the social fragmentation, or discontent that may threaten the dominant class, the system appeals to conspiracy or "counterfeits politics" to prevent any disruptive change. For better or worse, Americans believe erroneously they belong to a global superpower that should promote democracy worldwide. Terrorism, after all, reflects the own blame—as a mega-power—to inspire democracy in other "rogue states" or simply the resentment of others to be a prospering nation. In our chapter, we toy with the belief that terrorism is the repressed image of a declining-power which is gradually collapsing after the stock and market crisis in 2008 that accelerated a crisis that was already predicted.

REFERENCES

Achcar, G. (2015). *Clash of Barbarisms: The Making of the New World Disorder.* Abingdon: Routledge.

Altheide, D. (2017). *Terrorism and the Politics of Fear.* New York: Rowman & Littlefield.

Aly, A., & Green, L. (2010). Fear, Anxiety and the State of Terror. *Studies in Conflict & Terrorism, 33*(3), 268–281.
Amoore, L., & De Goede, M. (Eds.). (2008). *Risk and the War on Terror.* Abingdon: Routledge.
Aradau, C., & Van Munster, R. (2007). Governing Terrorism Through Risk: Taking Precautions, (Un)Knowing the Future. *European Journal of International Relations, 13*(1), 89–115.
Augé, M. (2002). *Diario de Guerra: el mundo después del 11 de Septiembre.* Barcelona: Gedisa.
Baudrillard, J. (2003). *The Spirit of Terrorism and Other Essays.* London: Verso.
Bellamy, A. J. (2006). *Just Wars: From Cicero to Iraq.* Cambridge: Polity Press.
Eid, M. (Ed.). (2014). *Exchanging Terrorism Oxygen for Media Airwaves: The Age of Terroredia: The Age of Terroredia.* Hershey: IGI Global.
Entman, R. M. (2003). Cascading Activation: Contesting the White House's Frame After 9/11. *Political Communication, 20*(4), 415–432.
Ericson, R., & Doyle, A. (2004). Catastrophe Risk, Insurance and Terrorism. *Economy and Society, 33*(2), 135–173.
Gray, C. S. (2007). *The Implications of Preemptive and Preventive War Doctrines: A Reconsideration.* Carlisle Barracks, PA: Army War College Strategic Studies Institute.
Griffin, D. R. (2004). *The New Pearl Harbor: Disturbing Questions About the Bush Administration and 9/11.* Northampton: Olive Branch Press.
Holloway, D. (2008). *9/11 and the War on Terror.* Edinburgh: Edinburgh University Press.
Hoskins, A., & O'Loughin, B. (2009). *Television and Terror: Conflicting Times and the Crisis of New Discourse.* New York: Palgrave Macmillan.
Howie, L. (2012). *Witnesses to Terror: Understanding the Meanings and Consequences of Terrorism.* New York: Springer.
Inboden, W. (2016). Reforming American Power: Civilian National Security Institutions in the Early Cold War and Beyond. In J. Suri & B. Valentino (Eds.), *Sustainable Security* (pp. 136–165). Oxford: Oxford University Press.
Kagan, R. (2004). *Of Paradise and Power: America and Europe in the New World Order.* New York: Vintage.
Kellner, D. (2003). *From 9/11 to Terror War: The Dangers of the Bush Legacy.* New York: Rowman & Littlefield.
Kellner, D. (2005). Baudrillard, Globalization and Terrorism: Some Comments on Recent Adventures of the Image and Spectacle on the Occasion of Baudrillard's 75th Birthday. *Baudrillard Studies, 2*(1), 1–15.
Kelman, D. (2012). *Counterfeit Politics: Secret Plots and Conspiracy Narratives in the Americas.* Lanham: Bucknell University Press.
Klein, N. (2007). *The Shock Doctrine: The Rise of Disaster Capitalism.* New York: Macmillan.

Korstanje, M. E. (2017). *Terrorism, Tourism and the End of Hospitality in the West*. Basingstoke: Palgrave Macmillan.
Korstanje, M. E. (2018). *Tracing Spikes in Fear and Narcissism in Western Democracies Since 9/11*. Oxford: Peter Lang.
Korstanje, M. E. (2019). *The Challenges of Democracy in the War on Terror*. Abingdon: Routledge.
MacGregor, D., & Zarembka, P. (2010). Marxism, Conspiracy, and 9/11. *Socialism and Democracy, 24*(2), 139–163.
Owens, J., & Pelizzo, R. (2010). Introduction: The Impact of the War on Terror on Executive-Legislative Relations: A Global Perspective. *The Journal of Legislative Studies, 15*(2–3), 119–146.
Pyszczynski, T., Solomon, S., & Greenberg, J. (2003). *In the Wake of 9/11: Rising Above the Terror*. Washington, DC: American Psychological Association.
Robin, C. (2004). *Fear: The History of a Political Idea*. Oxford: Oxford University Press.
Sahar, G. (2008). Patriotism, Attributions for the 9/11 Attacks, and Support for War: Then and Now. *Applied Social Psychology, 30*, 189–197.
Sheth, F. (2011). The War on Terror and Ontopolitics: Concerns with Foucault's Account of Race, Power Sovereignty. *Foucault Studies, 12*, 51–76.
Skoll, G. R. (2016). *Globalization of American Fear Culture: The Empire in the Twenty-First Century*. Basingstoke: Palgrave Macmillan.
Soyinka, W. (2005). *The Climate of Fear: The Quest of Dignity in a Dehumanized World*. New York: Random House.
Stiglitz, J. (2008). The $3 Trillion War. *New Perspectives Quarterly, 25*(2), 61–64.
Suri, J., & Valentino, B. (2016). *Sustainable Security: Rethinking American National Security Strategy*. Oxford: Oxford University Press.
Virilio, P. (2010). *University of Disaster*. Cambridge: Polity Press.
Weimann, G. (2005). The Theatre of Terror. *Journal of Aggression, Maltreatment & Trauma, 9*(3–4), 379–390.
Weisberg, J. (2008). *The Bush Tragedy*. New York: Random House Incorporated.
Weiss, T. G. (2006). R2P After 9/11 and the World Summit. *Wisconsin International Law Journal, 24*, 741.
Žižek, S. (2015). *The Universal Exception*. London: Bloomsbury.

CHAPTER 5

Tourism in the Days of Morbid Consumption

INTRODUCTION

Over the recent years, the academy has prioritized some themes which were associated directly or indirectly to development, sustainability, as well as those problems, emerged from real estate speculations in the main tourist destinations. In fact, capitalism found rapid ways of replication, and tourism became a formidable ally (Holden, 2003; Kampaxi, 2008; Lea, 1993; Mac-Beth, 2005). However, such an alliance has its costs. Recently, some voices questioned to what extent tourism, in neoliberal times, can be considered an ethical activity (Fennell, 2015; Garrod & Fennell, 2004; Hultsman, 1995; Malloy & Fennell, 1998). Part of the bibliography departs from the premise that tourists are in quest of pleasure maximization as the main criterion of attraction. What is more than important to discuss is the role played by the "Other" in this hedonist behavior?

Within philosophy, ethics are valorized as a form of enhancement which leads the self to progress. In tourism fields, for some reasons very hard to explain in this chapter, ethics was originally associated with the needs of protecting the interests of natives, or vulnerable stakeholders as women, children. Ethics are commonly understood as the system of codes that promotes the correct conduct escaping to the instrumentalization and material paradigms. While aborigines have been historically pressed to live in conditions of poverty through years, occupying a marginal position in the main economy, it is equally true that tourism offers a fertile ground to abandon poverty and slum, accelerating the conditions for fairer wealth distribution (Altman, 1989; Korstanje, 2012; Waitt, 1999).

Beyond European paternalism, which accompanied the inception and evolution of social anthropology, the fragility of the natives activates a sentiment of empathy from where tourism has historically evolved. The importance of understanding tourism as something else than a leisure activity inspired us to write this chapter, which explores not only the ebbs and flows of ethics in tourism but also the effects of neo-capitalism reproducing the economic background of poverty worldwide. To be more exact, one of the promises of the neoliberal discourse is associated to the opportunities of poor nations to improve their productive conditions through liberal trade and development. In second terms, the chapter scrutinizes whether modern tourism is—or not—an ethical activity debating the role of market and rationality as the main tenets of global consumption. Last days or bottom tourism exhibits a new drive to gaze the Other's pain, where the "Other" is neglected in favor of tourists' pleasure maximization. This idea cannot be understood, in anyway, without the inspection of the theory of development and the tourist gaze, which are the key elements for the commoditization of the "Other". Tourists, who seek to gaze spaces associated to trauma or mass-death, are not in search of empathy but replicate involuntarily the conditions of mourning. Let's cite the example of bottom-day tourism, a new tendency, where tourists visit tribes or spaces in danger of disappearance. They, tourists, do nothing to reverse the situation; rather they are indeed moved by the morbid taste of gazing "the end" of natives. In this way, tourists not only confirm their loyalties to the modern nation-states but also validate the idea that democracy and global capitalism as the best possible realms for them. As this chapter shows, this works as an ideological artifact whose origin stems from the old colonialism and the dark rationality of development theory. The system never corrects the causes of the tragedy but reproduces them to offer a dark spectacle, ready to be gazed upon!

Preliminary Debate

Dark tourism is often defined as a tendency to visit spaces of total obliteration or mass-death. Following Phillip Stone and Richard Sharpley (Stone & Sharpley, 2008), dark tourism evinces that the human drive orients to be emphatic with the Other's suffering. The self, aware of their complete finitude, interpret their own end through the Other's body. Other conceptions signal to the phenomenon as a cultural expression of heritage (Hartmann, 2014; Strange & Kempa, 2003) or as a mechanism of resiliency oriented to give an educational background (Cohen, 2011). Over the recent years, a great variety of publications appeared in the most important journals discussing different aspects of this morbid consumption which ranges from concentration camps (Lennon & Foley, 2000; Miles, 2002), natural or made-man disasters (Gotham, 2007; Hartnell, 2009), sites in bias of disappearance, post-conflict or disaster tourism (Chew & Jahari, 2014; Lisle, 2000; Robinson & Jarvie, 2008), prison tourism (Strange & Kempa, 2003; Wilson, 2011) and ghost castles or houses (Miles, 2015) and so forth. Although scholarship claims of the great dispersion this issue takes, at a first glimpse it is safe to say all sub-segments above mentioned follow the same dynamic (Korstanje, 2016b). Tourists in quest of these macabre places not only say that they need an authentic experience, they theorize over different mysteries concerning the human existence such as life, death, equality, democracy, prosperity and justice (to mention a few). The applied research, from where all these studies depart, focuses on the voice of interviewed tourists or visitors. Although the consulted tourists are a valid source of information, methodologically in some conditions they lie or simply protect their interests. This suggests that probably they are not interested in the Other's pain as they overtly utter. Secondly, and most important, there is a strong economic-centered view of these types of sites, without mentioning the fact that scholars prioritizes the consumption of heritage, and identity formation as key valuable elements of dark tourism. This nostalgia comes from old-dormant narratives framed in the European colonialism where the suffering native was seen ambiguously as an opportunity to protect its integrity and the sign of the cultural and biological supremacy of a (white) race over others. Doubtless to say the biological narratives that marked the inception of colonialism has set the pace to more subtle discourse but at the bottom the same logic remains: the non-Western "Other" is a subject of piety while a potential agent to be colonized, educated and ultimately controlled. These types of new "morbid forms of consumption" reveal two important things. On one hand, the consumption of death is not limited to tourism; rather it is present in all the media. This means that we are next to a new stage of capitalism where death plays a leading role mediating between citizens and their institutions. On the other hand, by gazing at the tragedies involving others remind us how important our system of beliefs is, how significant democracy and Western civilization are. The point of departure of this reasoning sheds light on the *rationality of development* which places *progress* as the dividing line between the salved and the doomed groups. As discussed in earlier chapters, this viewpoint illustrates the nature of a radical puritanism which was founded in an apocalyptic cosmology of the world. We need to visit those landscapes before their disappearance because the big fish (capitalism) eats the weaker fish. Conceiving capitalism as a taken-for-granted institution, we accept involuntarily there is nothing beyond its borders. In this way, our critical thinking is domesticated toward a solipsist world where the man is a wolf to another man.

The Rationality of Development

The question whether tourism can be considered a curse or a vehicle toward development was originally addressed by Emanuel de Kadt. In his book, *Tourism: Passport to Development?*, he writes that some nations are culturally open to embrace development while others are not. In a nutshell, those cultures historically subject to a past of slavery or exploitation has fewer probabilities to boost their economies than others. The adoption of authoritarian cultural values impedes of the necessary levels of freedom to avoid extractive institutions, which are commonly based on speculation and corruption. The lack of democracy, in a long term, develops a pathological cultural character plagued by social conflict, pillage, corruption and other social maladies. For de Kadt, tourism is only a direct result of the prolonged state of democracy that characterizes the US and Western Europe (de Kadt, 1979). In de Kadt's account, as well as many theorists of development, the problem of development is not constituted in the content of programs, unless by the fact that some cultures are symbolically determined to accept authoritarian values that impede the revitalization of democratic institutions. Those countries, which kept a past of violence, genocide or slavery, have lesser probabilities to embrace development as the main option than other nations which forged a stable democracy. This discourse, which was reinforced during neoliberalism, not only resonated in tourism fields during the 1990s but also persists up to date. The current narratives emulated by academy envisage tourism as the main instrument toward development, revitalizing history and tradition and gaining further profits for the community. Since authorities need to foster stable economies for gaining re-elections, it is equally true that tourism will alleviate poverty endorsing further legitimacy to politicians.

Jose Osmar Fonteles criticizes the idea of development as a wonderland where all material deprivations are addressed. Development not only is far from starting a genuine growth but under some conditions affects the locals. The life of residents is often altered by the arrival of tourists as well as the environmental problems the activity left. Under the dichotomy of rationality, what development exhibits are the lack of any rational basis simply because profits are expatriated to the US and Northern Europe (precisely places where investors and tourists come from), while territory and its resources are seriously placed in jeopardy. Fonteles (2004) acknowledges that beyond the popularity of sustainable tourism and development, pollution is considerably higher today than other times. In order to prevent real estate speculative policies, governments should directly intervene to protect the interests of all stakeholders.

The rationality of development seems not to be pretty different to the ideals of Roman Empire. Ancient Romans believed not only in their supremacy but in the power of trade. They believed that the uncivilized cultures may improve their conditions by dealing with an established civilization. This helped Romans not only in deploying their armies—constructing ways that allowed the travels of thousands of tourists—but also in extracting basic resources to be reimported in the form of stylized merchandises (Korstanje, 2009). The same archetype, paragraphing Anthony Pagden, was emulated by Spanish conquistadors but also by the Anglo-Saxons in the Americas. The conquest starts from a cosmology, a belief of the external World where the alterity is imagined, constructed and subordinated to certain mainstream cultural values. Through the articulation of ethnocentrism, the civilizing process says what is desired and what is not. The concept of rationality divides the world between the desired citizens and those who are not accepted as humans (Pagden, 1995). In the threshold of time, as Pagden shows, the Westphalian nation-state was centered on two main values, free trade and mobilities. In consequence, Occident developed an uncanny vision of the Otherness, sometimes frightened as a savage enemy, while others as "a noble Savage" (Korstanje, 2017).

The rationality of development today, like the nostalgia for heritage by the side of the anthropology, is determined by the needs of experiencing authentic experiences, or gazing at "the Other" as a mirror. However, as John Urry puts it, far from being ad-hoc, the modern gaze is previously conditioned by economic and social forces that shape it. Urry, in fact, coins the term "tourist gaze" as an attempt to describe a visual attachment to possess what is being watched. Though he glossed over the fields of ethics, the tourist gaze encompasses three constitutive aspects: *the re-enchantment of consumption, time-space dimensions* and *visualization of performing arts*. Therefore, travels re-symbolize spaces at the time they are transformed into commodities. The potential consumers, tourists, are bombarded by images: advertising and visual stimuli channeling them to interpret landscapes in specific ways. The decentralized control of space is accompanied by solicitation of credit from international banking apparatuses. Such national policies create a dependency of tourist-destination state to global capital. Late modernity deregulates finance, and thereby introduces an uncontrolled flux of capital which re-draws the contours of a globalized world. As J. Urry (2002) notes, the global modernity is based on the monopoly of signs which are produced, commercialized and decentralized to be individually consumed. In Urry's account, landscapes produce gazes to be visually consumed, while individually these landscapes are selected according to each particular biography. He was a pioneer among scholars who thought that we are in conditions to forge a discipline aimed at exploring the paradigm of mobilities. Travels often, Urry acknowledges, activate some specific-contextualized forms of gazes broadly classified according to a much deeper cultural matrix, which gives very well meaning to the gazed object. After all, we move through symbolic articulations which give meaning to events. This point reminds that any movement, of course, is a type of negotiation, a rite where interaction pivots (Lash & Urry, 1994; Urry, 2002). It is a memorable debate between Urry and MacCannell. Precisely, in the opposite direction, Dean MacCannell interrogates in his book, which is entitled the *Ethics of Sightseeing*, furtherly on the intersection of modernity, consumption and tourism. The problem lies neither in the individual agency nor in its capacity to interpret the landscape but it stems from the epistemology of capitalism which ponders consumption as something valued by all members of society. While watching represents a hermeneutical interpretation, connecting peoples with landscapes, and experiences, the derived meaning of events are not chaotically organized in social imaginary. There is a meta-discourse which precedes the patterns, horizons and constellations that operate in the individual behavior. To what extent the message is correctly deciphered appears to be the key issue of any social scientist. Unlike other texts, *the Ethics of Sightseeing* is unique because the debate is posed in the direction of philosophy. The history witnessed how the sacred-life has set the pace to more secular forms. If the figure of the Totem, as a system of beliefs, plays a leading role in organizing the life of society, tourism plays an analogous role in modern societies. In the MacCannell's conception, West has advanced in the trace of technology imposing to the world a new secular view of the environment and the connection with the otherness. The sacred-space that characterized the gaze of aborigines, today, has been replaced by a new one more elaborated, secular and fictional sightseeing: *tourism*. What is important to discuss here seems to be the connection of tourists with social symbolism that mediates between self and consumption. Money serves a mediator connecting all dispersed consumers. In late modernity, consumption and mobility pave the ways for the advance of alienation and depersonalization. Since tourism commoditizes others by the introduction of sightseeing, MacCannell is not enthusiastic about the fact; tourism would help communities to abandon poverty and alienation (MacCannell, 2011). In this way, tourism serves as an ideological instrument that subverts the possibilities of workers to confront the ruling elite, and in so doing, it increases the poverty. The exploitation of the lords of industrialism, the Marxian texts indicate, is hidden by the action of ideology. Tourism empties spaces cannibalizing the alterity, while the culture is exchanged and commoditized to the taste of First World tourists. Given the problem in this term, the profits mediate between citizens increasing the probabilities of conflict, even war.

These concerns are in consonance with the seminal book authored by Comaroff and Comaroff, *Ethnicity INC*. Beyond the promises of the globalized world, per the argument readers will find in this text, there is an underlying logic of commoditization and consequent exploitation by which human beings, their cultures and traditions become business enterprises. Since anthropology as an academic discipline emerged as an extreme concern in founding parents to the disappearance of non-Western cultures, this sentiment has taken the opposite direction. In our days, aboriginals appeal to reinforce their own differences to be sold to the international segment of travelers and tourists. In perspective, cultural tourism is one of the most growing industries in the world. Ethnicity, in this vein, set the pace to a new type of cultural consumption fabricated from outside to regulate emotions. The term "empowerment" as it has been formulated by the specialized literature is defined as strategy followed by local actors to improve economic and social conditions by means of proactive participation and commitment. When aborigines adopt "empowerment" simply because they know something special can be offered to the international Western consumer, their culture is recycled as a commodity. This new type of identity, though more flexible, objectifies the native to the extent to its needs are enslaved to a fabricated past. Basically, cultural tourism not only evokes a vibrant past which does not exist but also confers to local communities the legal mechanism for launching to self-representation. The value of aboriginal cultures is conditioned by those features that legitimate the West supremacy. Aboriginals may say something if this discourse can be commercialized. This alludes to a much deeper process of alienation where cultures are disclosed from their original roots. In doing so, the culture is sold attending only to the interests of consumers. The enthusiasm and leading role of aborigines as cultural managers blur the conflictive relations of Fourth World and States but creating new ones. Here two assumptions should be done.

On one hand, tourism disposes from cultural protection to re-draw the geography of the world. On the other hand, native constructs their sentiment of belonging in view of what tourists want to hear and see. The merit of this work consists in reminding that this trend not only blurs the boundaries between past and present but also impose new economies based on ethnic-merchandise where the production never ends. The classic rules of economy teach us that the rise of demands entails a decline in the production. Needless to say, this does not happen with ethnic-merchandise. The much demand for cultural consumption, the better for production; that way, the destination never declines in extractives. One of the problems of heritage and tourism are expressed in the violence exerted by states when aborigines reject the possibility to be taxed. While development brings some benefits to local communities, it also triggers a situation of political crisis that very well leads society into a civil war—of course unless duly and timely regulated (Comaroff & Comaroff, 2009).

Tourism, Colonialism and Development

The intersection of development and colonial order was brilliantly addressed by P. McMichael (2016) in the book *Development and Social Change*. His main thesis is that Europe, by the introduction of "colonialism", established an ideological background for legitimizing their submissions to its overseas colonies. The exploitation of the non-European "Others" had a pervasive nature. Sooner, aborigines realized the double moral standards of the colonial order. While cruelty, submission, and violence were applied in the colonies, in the core democracy prevailed as a valid system of government. This opens the doorstep to the process of "decolonization", where thousands of peripheral voices claimed to access the same rights "the democracy of their white lords" declared. McMichael explains that imperial powers alluded to the theory of "development" to maintain the dependency between the center and its periphery. The WWII end conjoined to Truman's administration led the US to implement a wide range of credit system to save the world from communism. This program mushroomed to become the development theory. As a mega-project, the theory of development was coined in 1940 and lasted till 1970. It not only created a food dependency but also accelerated the numbers of slum-dwellers and poverty in the peripheral countries. In order for remaking the old division of labor, imperial powers induced "Third World" to accept international loans, which were used to industrialize their economies. At the time, under-developing nations adopted capital-intense methods in agriculture ruining the condition of small farmers, who migrated to urban areas; the US and Europe exported industrialized products. It was unfortunate that the effects of development were far from being the ideals for Africa. The old boundaries of ethnicities the first colonial powers found were never honored once WWII finalized. Many human groups were forced to live together within fabricated borders that delineated the sovereignty of the new-born nation-states. This resulted in a lot of ethnic cleansing, conflicts, and warfare that obscured the original ends of financial aid programs issued by the IMF or World Bank. Undoubtedly, the inconsistencies of the World Bank in administering the development-related programs not only were admitted but also it woke up some nationalist reactions in the non-aligned countries. To restore the order, a new supermarket revolution surfaced: *globalization.*

As the previous argument given, McMichael argues convincingly that globalization succeeded in expanding thanks to the lack of protective barriers of the Third World where the capital investors were welcomed. This, in consequence, provoked two alarming situations. An increase in the unemployment and the decline of unionization in the North was accompanied by the arrival of international business corporations seduced by the low cost of workers in the global South. The doctrine of "free enterprise" was discursively presented as a superior ladder in the evolutionary process. Each state should adopt a specialized role in a much wider "world factory" where some provide the raw materials and others elaborated products. These trends which characterize the 1990s created a new asymmetry between skilled (located in the First World) and under-skilled human resources (situated in the periphery). The recession produced by oil embargo pressed the First World to generate a massive influx of money toward the Third World, although to be carefully vetted by two organizations, GATT and WTO. Both curtailed the protective measure of local economies by consolidating a new model which combined the reduced public capacity with the needs of governance. If nationalism showed the importance of nation-state to protect the citizen from market's arbitrariness, now neoliberalism focused on the inefficiency of public administration to regulate the economy.

> In short, the making of a free trade regime reconstructed food security as a market relation, privileging and protecting corporate agriculture and placing small farmers at a comparative disadvantage. Food security would now be governed through the market by corporate, rather than social criteria. (p. 136)

On one hand, the World Trade Organization was relentless by charging or applying sanctions over those countries that affected the new system of import-export. Less interested in freeing trade than in consolidating their hegemony, main powers prompted the discourse that Third World had not the right toward "self-sufficiency" anymore. On the other hand, globalization accelerated the accumulation of profits (beyond the boundaries and control of nation-states) but enlarging the levels of poverty as never before. Per some calculations, UN declared that only 20% of the world population is situated within 20% of the richest people, whereas the rest is facing serious economic problems. It is important not to lose the sight that almost 1 billion peoples are living in slums. Many of the First World tourists forget this reality conceiving mobilities as a universal value (Korstanje, 2018). This raises the question how does ideology really work?

In earlier research, we explained that the limitations of development theory rest not only in its ideological nature but also in how the language and grammar is politically manipulated. Basically, white-elite mark ethnicities to control them whereas—in so doing—they avoid any type of mark themselves. To set an example, the term "ethno-tourism" or "cultural tourism" is daily conceived to denote visits to aboriginal reservoirs, while a travel to Chicago (Illinois) does not receive the same label. The needs of marking others entail being avoided to be marked. This means that the core of our language contains "a strong paternalism" which surmises Europeans as superior to other values. With this in mind, the theory of development follows an ideological connotation because of two main reasons. The subordinated role of aborigines impedes the necessary platforms to break their dependency with respect to capitalist nation-state but, secondly, the theory of development introduces European cultural values which are linked to control and rationality, as the best of possible worlds. Being developed or not developed is externally considered according to Western coefficient and—of course—a rationality aboriginals do not share. Given this argument, the exegetes of development are blind to see that the failure of the theory is not explained by the incapacity of the Western rationality to be applied in non-Western groups but as cultural pathologies of natives (Korstanje, 2012). Is tourism an ethical activity?

THINKING ETHICS IN TOURISM

Countless studies have focused on the commercial nature of hospitality (Alvarez & Korzay, 2008; Capriello & Rotherham, 2008; Castaño, Moreno, & Crego, 2006; Coronado, 2008; Franch et al., 2008; Heuman, 2005; Kastenholz & Lopez de Almeida, 2008; Lashley & Morrison, 2001; Lau & Mckercher, 2006; Lynch, 2005; McNaughton, 2006; Nadeau et al., 2008; Santana, 2006; Santos Filho, 2008; Toribio Camargo, Castellá Cordoba, & Serrano Orquin, 2005), while less attention was given to the ethical dilemma given by the profit-oriented paradigm. At a closer look, hospitality may be only ensured when guests have material resources to pay for it. Neither migrants nor any other undesired guests are accepted by nation-state beyond the paradigm of gift-exchange (Korstanje, 2016a; Korstanje & Olsen, 2011). On the other hand, the process of touristification leads local communities toward social benefits but unfortunately compromising some natural resources when real estate is uncontrolled. Is the quest for profits (which is the symbol of the economic-centered paradigm in tourism) unethical by nature?

Some empirical publications will help in debating this point. The research of Franch et al. explores the Global Warming as a threat to Alpine Areas (the Dolomites area) as well as the urgency of innovative strategies in order for managing unexpected effects that accelerate a profound change in way hosts are symbolically interconnected with their landscapes. The Global Warming impinges on destroying not only the environment by means of snow reductions but also the different industries directly or indirectly linked to tourism and hospitality and of course to winter season exploitation; Franch et al. argue that diversification of business and technology are more than needed to overcome these new pitfalls; by analyzing the testimonies of 16 cableway associations in Trentino, the present investigation revealed deficiencies on radical innovation and behavioral adaptation to global warming with the exemption of investments in artificial snow-making (Franch et al., 2008). Policy-makers should take note of the degree of dependency the community developed with respect to tourism. Ohe (2010) observes that the performance of rural destinations can alleviate poverty at a short run but create a long-run dependence when profitability is pondered as the only requisite. Judging ethics according to context can give answers in some extreme cases, but it is not recommendable for all cases.

Is Rationality the Heart of the Market?

An authoritative voice in the economy, Robert Sugden, lecturer at University of East Anglia UK, agrees that society evolves from a spontaneous order early conditioned by previous negotiations to a state of further stability where individualism should be sacrificed. In fact, the quest of goal-oriented maximization is left behind to ensure shared conventions. Among frights and nightmares citizens face, the behavior is oriented to avoid "chaos" and disorder. The society is united because lay citizens endorse solidarity to their social institutions, where each side commits to respect the rules only under the premise the other does the same (Sugden, 2005). This is the position from where virtue operates in day-to-day contexts. As stated, in some respect, virtue is defined by Nancy Snow as "enduring character traits incorporated practical reason, appropriate motivation, and affect, and manifested in cross situationally consistent behaviour" (Snow, 2010: 117).

As the previous argument given, she contends that behavior takes two different dynamics: *habitual and deliberative*. While habitual virtuous acts come from any previous deliberation since the act has been routinized, deliberative sub-type influenced people to make a decision. Though, in philosophy, situationists are consistent with their criticism to the theory of traits, they ignore the fact that behavior is a multifaceted result where motivations, affects and history are organized by the mind. Each person may adjust their own desires in hoping to reach a virtuous result. The development of virtue is in sync with the cultivation of ethical decisions. In times of globalization, we place ethics before the world of business— not for ethical purposes but only to enhance profits. This represents the case of objection presented by Slavoj Žižek about MacDonald and other fast-food corporations which sell their products enmeshed in a charitable project where consumers help to struggle against poverty in Africa, Latin America or Asia (Žižek, 2008).

What is worse, further attention should be given to slum tourism, which consists in practices overtly aimed at visiting spaces of mass poverty to get "authentic experience" with the others, but in so doing, visitors takes asunder from real ethics. A recent and seminal work edited by Clare Weeden and Karla Boluk (*Managing Ethical Consumption in Tourism*, 2014) shows that there is a clear need to rethink the classic definition of tourism. Many emerging sub-segments defied the connotation of mass-tourism or even sun-and-beach products over recent years. From dark to slum tourism, the notion of beautiness is experiencing substantial changes. For specialists as Weeden and Boluk, ethics cannot be applied in all contexts because what turns out legal in some nations are not in others. Despite the numerous studies focusing on how tourism stimulates consumption are far from taking ethics as a serious option. Tourist destinations, as Weeden and Boluk examine, embrace some ethical values only when they can be successfully commercialized to earn further advantages in a globalized and ever-changing market. This begs a thorny question: is the commercial nature of tourism divorced from ethics?

One of the aspects that undermines the role of ethics in tourism fields is the belief that responsibility relates to the individual stance (optional) to gain some certification schemes. In this vein, we are not legally pressed to adopt ethics in our professional life, unless it grants additional profits. What this book puts in straights is how beyond the discourse of "responsible tourism" there is an ambiguous position where our moral duties are not a question of law but of election. A second more epistemological reason lies in the fact, studies in tourism ethics are scattered or published in humanitarian journals more associated to philosophy, sociology or anthropology instead of coordinating all produced knowledge in only one magazine or platform.

> What is unique about ethics is that more than any of these other fields of inquiry, it gets to the job of illuminating the tension that exists between what is good and bad, right and wrong, and authentic and inauthentic in tourism. It does this by forcing us to examine the human condition first, knowing something of ourselves, as a way to understanding tourism. (Preface, Fennell, 2015: XVI)

From a broader sense, Weeden and Boluk remind though sustainability has recently monopolized a great variety of research which focuses on the failure in developmental programs or negative effects of the industry in local communities, tourism-related policies are profit-centered. The questions whether the sphere of ethics is dissociated from consumption explain why tourists are often insensitive to Other's sufferings. The system never corrects the inequalities but paves the ways for appreciating them as a show.

Another interesting point of convergence in this discussion is presented by Bianca Freire Medeiros (2015), and Rodanthi Tzanelli. While the former in her book (*Touring Poverty*) rationalizes tourism as a need to watch something different alluding to the escapement, the latter sets an interesting debate connecting slumming with colonialism. The postmodern tourists found in the Other's pain the ideological answer to avoid the responsibilities of their nation-states in the age of colonialism. Instead, Freire Medeiros rejects the possibilities slum tourism really helps others to get away from Favelas and slums, even the needs of selling poverty poses a much deeper ethical dilemma; if poverty becomes the commodity of tourism, it will be never eradicated. Since commoditized to gain attractiveness, poverty idealizes the miserable conditions of exploitation favelados are facing daily. Beyond the ethical discussion to what extent slum tourism assists residents to alleviate the symptoms of poverty, Freire Medeiros writes that in the Favelas the nets of interactions lead to reify tourism as a mechanism to improve residents' lives but in doing so, it paradoxically produces poverty. To put the same in other terms, in the theory of classical economy, commodities are the vital part of merchandise production. In slum tourism, the infrastructure, transport, restaurants, tour operators, tour guides and every service is unfortunately based on "the pauperization" of natives. Hence, the poverty becomes a commodity which far from being corrected is finally replicated.

Rather, Tzanelli's insight is not provided by fieldwork (as the book presented by Medeiros) but from a visual ethnography on the cinematic representation of slums in movies entertainment industry. Her main outcome exhibits how the chief ideals and narratives of colonialism still operate through the plots of cinema industry. Tourists who look for slums not only reinforce an old romantic supremacy but also ascribe to the cultural ideals of global capitalism (Tzanelli, 2016).

The rise of demand for the consumption of slum tourism, which means the attractions for gazing spaces of poverty, misery and pauperization, seems to be conducive to "the logic of social Darwinism" (Korstanje, 2015), where few selected people see how the rest simply perish. Over the last decades, slum tourism has not deterred the advance of poverty, rather replicated it. From colonialism to the industrial revolution, Zygmunt Bauman (2000) adds, the question of poverty was replicated in order for the status quo to be maintained. While the financial global elite visit the exotic archaeological ruins, thousands of migrants are disciplined in their arrival to Europe. The term "disciplined" here is employed to denote how overseas migrant workforce that arrived in the US at the end of the nineteenth century was subject to strict patterns of re-education and control. Some cultural values such as self-determination, development and freedom were the conceptual platforms for the expansion of modern capitalism.

In a recently released book, we used the term "Thana-capitalism" to denote a new stage of production in globalized economies, where the society of risk as it was imagined by the cultural sociologists sets the pace to a facet of consumption where death plays a crucial role. News of disasters, novels, realities, and other cultural entertainment programs reveal not only modern citizens developed a strange aversion to death, but they feel happiness from others' suffering because it is the only way to deride from death. In a climate of extreme competition legalized by social Darwinism, individualism prevails through the lens of consumption. In times of Thana-capitalism, there is no place for ethics, since the role of empathy is emptied. One of the best examples to explain this further is the plot of *Hunger Games*, where capitol organizes the Games. In these spectacles of death, participants not only do not cooperate with others, they exaggerate their real probabilities to defeat. Only one can be the winner, and of course, the winner takes everything! This suggests that those material asymmetries which ideologically legitimize the interests of elite are internalized by citizens as a natural-like situation (Korstanje, 2016b). The patterns of gazing and consumption are changing toward versions that are more morbid. The turn of the century witnessed not only the rise and expansion of global risks as terrorism, virus outbreaks or natural disasters but the productive system, far from correcting their causes, recreates the conditions to offer "a spectacle" that keeps often the modern audiences entertained. This position leads to a philosophical problem because since the causes of the events are ideologically tergiversated, the probabilities to experience a new disturbing disaster turn higher. In the conclusion, we explain the main psychological profile of those tourists who seek these types of dark experiences, trying to construct a typology which is not exhaustive but gives illustrations for next approaches.

Conclusion

Throughout this chapter, we ignited a hot debate on the role of globalization as a chief agent oriented to connect dissimilar economies into an all-encompassing system. Although the chapter is abstract, so to speak not empirically based on, we laid the foundations toward a new sociological idea: the mutation of the risk society into a new version where the consumption of death is the main commodity. The question of whether tourism should be considered ethical or not still remains open. We feel we are not in a position to affirm tourism is unethical but in days of Thana-capitalism, the suffering pivoted as the main commodity not only that helps structuring social institutions, but the necessary mediator between lay people and their states. The commoditization of tragedy transforms the sensitivity of people making them more hedonists. In addition, far from being altered the conditions of exploitation inaugurated by neo-globalization after 1990s have been enlarged. In order for the system not to collapse, the necessary undesired effects such as violence, riots, poverty and suffering are being recycled as the main criterion of attraction for the postmodern tourist destinations. Nonetheless, this is not limited to tourism but can be traced back to other media entertainment too. This is exactly the point we do not stop consuming news containing images that can affect viewers' sensibility, but indeed we are eager to do it. As MacCannell observed, the limitations of ethics to be framed within consumption rest on the belief that there is a long-simmering divorce between what people wish and what they should do. Since modern tourism is a profitable industry oriented to satisfy the hedonist drives which lead to a "pleasure maximization logic", ethical rules only can be externally imposed by the state. In times of our grandparents, holidays represented a sacred space to be enjoyed in paradisiacal landscapes or sunbathing at a named beach. The chief concept to understand with regards to these holiday-makers in terms of Krippendorf was not only the needs of escapement but also the concept of an apollonian sense of beautiness. This world has gone forever, and this gazing sets the pace to new connotation associated to tragedies, mass death, poverty and all types of human miseries. These death-seekers are in quest of others' suffering in order to confirm these privilege positions as First World consumers. Although a complete psychological profile needs further explorations, the following scheme will serve to help understand the psychology of death-seekers:

- They feel prone to witness events that do not involve them directly. This type of vicarious empathy in rare occasions can be crystalized in real assistance to others.
- They valorize present times than past because they keep an ethnocentric viewpoint of events.
- Death-seekers only embrace heritage to understand this time is the best of the possible realms.
- Dogmatic in their personality, they do not understand reality unless by their earlier cognitive frames.
- Sites of mass-death, disaster or suffering (Thana-tourism) are often selected as the primary destinations for visiting in holidays.
- Since they are special, death consumers feel they have the right to interact with others well-skilled like them.
- Death-seekers support social Darwinism where the survival of strongest is the main cultural value.
- Consuming others suffering they feel special, superior or more important.
- Frightful personalities that think the world is a dangerous place.
- Death-seekers entertain witnessing how others struggle. Very open to mythical conflagrations as goodness against evilness, they symbolically associate death to "condemnation". For them, the correct persons should not die.
- Pathological problems to understand death.
- Regardless the political affiliation, they embrace "counterfeit politics", or the theories of conspiracy.

Last but not least, the concept of Thana-capitalism coheres with previous attempts to describe the expansion of narcissist culture in the early works published by C. Lasch (1991) or the theory of other-oriented typology as it was formulated by David Riesman (2001) in his project, *A Lonely Crowd*. Though the concept of narcissism is still present in the current society of consumption, what Lasch and Riesman left behind was the role of death as a perfect symbolizer of the exclusion narcissism generates. Engaged in forms of spectacle, Thana-capitalism recycles the conditions which may be helpful for a social change into reified products which are widely witnessed by the spectorship through the lens of TV and media. The point will be explained in the next chapter.

References

Altman, J. (1989). Tourism Dilemmas for Aboriginal Australians. *Annals of Tourism Research, 16*(4), 456–476.
Alvarez, M., & Korzay, M. (2008). Influence of Politics and Media in the Perceptions of Turkey as a Tourism Destination. *Tourism Review, 63*(2), 38–46.
Bauman, Z. (2000). *Globalization: The Human Consequences*. New York: Columbia University Press.
Capriello, A., & Rotherham, I. (2008). Farm Attraction, Networks and Destination Development: A Case Study of Sussex, England. *Tourism Review, 63*(2), 59–71.
Castaño, J. M., Moreno, A., & Crego, A. (2006). Factores psicosociales y formación de imágenes en el turismo urbano: un estudio de caso sobre Madrid. *Pasos, 4*(3), 287–299.
Chew, E. Y. T., & Jahari, S. A. (2014). Destination Image as a Mediator Between Perceived Risks and Revisit Intention: A Case of Post-Disaster Japan. *Tourism Management, 40*, 382–393.
Cohen, E. H. (2011). Educational Dark Tourism at an In Populo Site: The Holocaust Museum in Jerusalem. *Annals of Tourism Research, 38*(1), 193–209.
Comaroff, J. L., & Comaroff, J. (2009). *Ethnicity, Inc.* Chicago: University of Chicago Press.
Coronado, G. (2008). Insurgencia y Turismo: reflexiones sobre el impacto del turista politizado en Chiapas. *Pasos, 6*(1), 53–68.
de Kadt, E. (1979). *Tourism Passport to Development? Perspectives on the Social and Cultural Effects of Tourism in Developing Countries*. Oxford: Oxford University Press.
Fennell, D. A. (2015). Ethics in Tourism. In G. Moscardo & P. Benkendorff (Eds.), *Education for Sustainability in Tourism* (pp. 45–57). Berlin: Springer.
Fonteles, J. O. (2004). *Tourism and Socio-Enviromental Factors*. Sao Paulo: El Aleph.
Franch, M., et al. (2008). 4l Tourism (Landscape, Leisure, Learning and Limit): Responding to New Motivations and Expectations of Tourist to Improve the Competitiveness of Alpine Destination in a Sustainable Way. *Tourism Review, 63*(1), 4–14.
Garrod, B., & Fennell, D. A. (2004). An Analysis of Whalewatching Codes of Conduct. *Annals of Tourism Research, 31*(2), 334–352.
Gotham, K. F. (2007). (Re)Branding the Big Easy: Tourism Rebuilding in Post-Katrina New Orleans. *Urban Affairs Review, 42*(6), 823–850.
Hartmann, R. (2014). Dark Tourism, Thanatourism, and Dissonance in Heritage Tourism Management: New Directions in Contemporary Tourism Research. *Journal of Heritage Tourism, 9*(2), 166–182.
Hartnell, A. (2009). Katrina Tourism and a Tale of Two Cities: Visualizing Race and Class in New Orleans. *American Quarterly, 61*(3), 723–747.

Heuman, D. (2005). Hospitality and Reciprocity Working Tourist in Dominica. *Annals of Tourism Research, 32*(2), 407–418.

Holden, A. (2003). In Need of New Environmental Ethics for Tourism? *Annals of Tourism Research, 30*(1), 94–108.

Hultsman, J. (1995). Just Tourism: An Ethical Framework. *Annals of Tourism Research, 22*(3), 553–567.

Kampaxi, O. (2008). Codes of Ethics in Tourism: Practices, Theory and Synthesis. *Annals of Tourism Research, 35*(2), 607–608.

Kastenholz, E., & Lopez de Almeida, A. (2008). Seasonality in Rural Tourism – The Case of North Portugal. *Tourism Review, 63*(2), 5–15.

Korstanje, M. E. (2009). Reconsidering the Roots of Event Management: Leisure in Ancient Rome. *Event Management, 13*(3), 197–203.

Korstanje, M. E. (2012). Reconsidering Cultural Tourism: An Anthropologist's Perspective. *Journal of Heritage Tourism, 7*(2), 179–184.

Korstanje, M. E. (2015). *A Difficult World, Examining the Roots of Capitalism*. New York: Nova Science Publishers.

Korstanje, M. E. (2016a). Pensando la hospitalidad. *Revista de Sociales y Jurídicas, 11*, 208–214.

Korstanje, M. E. (2016b). *The Rise of Thana Capitalism and Tourism*. Abingdon: Routledge.

Korstanje, M. E. (2017). *Terrorism, Tourism and the End of Hospitality in the West*. New York: Palgrave Macmillan.

Korstanje, M. E. (2018). *The Mobilities Paradox: A Critical Analysis*. Cheltenham: Edward Elgar.

Korstanje, M. E., & Olsen, D. H. (2011). The Discourse of Risk in Horror Movies Post 9/11: Hospitality and Hostility in Perspective. *International Journal of Tourism Anthropology, 1*(3–4), 304–317.

Lasch, C. (1991). *The Culture of Narcissism: American Life in an Age of Diminishing Expectations*. New York: W. W. Norton & Company.

Lash, S., & Urry, J. (1994). *Economies of Signs and Space*. London: Sage.

Lashley, C., & Morrison, A. (2001). *In the Search of Hospitality*. London: Butterworth Heinmann.

Lau, G., & Mckercher, B. (2006). Understanding Tourist Movement Pattern in a Destination: A GIS Approach. *Tourism and Hospitality Research, 7*(1), 39–49.

Lea, J. (1993). Tourism Development Ethics in the Third World. *Annals of Tourism Research, 20*(4), 701–715.

Lennon, J. J., & Foley, M. (2000). *Dark Tourism*. London: Cengage Learning EMEA.

Lisle, D. (2000). Consuming Danger: Reimagining the War/Tourism Divide. *Alternatives, 25*(1), 91–116.

Lynch, P. (2005). Sociological Impressionism in a Hospitality Context. *Annals of Tourism Research, 32*(3), 527–548.

Mac-Beth, J. (2005). Towards an Ethics Platform in Tourism. *Annals of Tourism Research*, 32(4), 962–984.
MacCannell, D. (2011). *The Ethics of Sightseeing. Dean MacCannell.* Los Angeles: University of California Press.
Malloy, D. C., & Fennell, D. A. (1998). Codes of Ethics and Tourism: An Exploratory Content Analysis. *Tourism Management*, 19(5), 453–461.
McMichael, P. (2016). *Development and Social Change: A Global Perspective.* Thousand Oaks: Sage Publications.
McNaughton, D. (2006). The Host as Uninvited Guest: Hospitality, Violence and Tourism. *Annals of Tourism Research*, 33(3), 645–665.
Medeiros, B. F. (2015). *Touring Poverty.* Abingdon: Routledge.
Miles, T. (2015). *Tales from the Haunted South: Dark Tourism and Memories of Slavery from the Civil War Era.* Chapel Hill: University of North Carolina Press.
Miles, W. F. (2002). Auschwitz: Museum Interpretation and Darker Tourism. *Annals of Tourism Research*, 29(4), 1175–1178.
Nadeau, J., et al. (2008). Destination in a Country Image Context. *Annals of Tourism Research*, 35(1), 84–106.
Ohe, Y. (2010). Barriers to Change in Rural Tourism. In P. Keller & T. Bieger (Eds.), *Managing Change in Tourism: Creating Opportunities – Overcoming Obstacles* (pp. 31–45). Berlin: Erich Schmidt Verlag.
Pagden, A. (1995). *Lords of All the World. Ideologies of Empire in Spain, Britain and France c. 1500–c.1800* (p. 64). New Haven: Yale University Press.
Riesman, D. (2001). *The Lonely Crowd.* New Haven: Yale University Press.
Robinson, L., & Jarvie, J. K. (2008). Post-Disaster Community Tourism Recovery: The Tsunami and Arugam Bay, Sri Lanka. *Disasters*, 32(4), 631–645.
Santana, A. (2006). *Antropología y Turismo.* Barcelona: Ariel Ed.
Santos Filho, J. (2008). Hospitalidade no Brasil Imperio: a visao o naturalista George Gardner. *Revista Brasileira de Pesquisa em turismo*, 2(2), 3–19.
Snow, N. E. (2010). *Virtue as Social Intelligence: An Empirically Grounded Theory.* Abingdon: Routledge.
Stone, P., & Sharpley, R. (2008). Consuming Dark Tourism: A Thanatological Perspective. *Annals of Tourism Research*, 35(2), 574–595.
Strange, C., & Kempa, M. (2003). Shades of Dark Tourism: Alcatraz and Robben Island. *Annals of Tourism Research*, 30(2), 386–405.
Sugden, R. (2005). *The Economics of Rights, Cooperation and Welfare.* Hampshire: Palgrave Macmillan.
Toribio Camargo, A., Castella Cordoba, P., & Serrano Orquin, I. (2005). Determinación de las preferencias de los clientes internacionales para la practica del turismo rural en la Republica de Cuba. *Pasos*, 3(2), 283–295.
Tzanelli, R. (2016). *Thanatourism and Cinematic Representations of Risk.* Abingdon: Routledge.

Urry, J. (2002). *The Tourist Gaze*. London: Sage.
Waitt, G. (1999). Naturalizing the 'Primitive': A Critique of Marketing Australia's Indigenous Peoples as 'Hunter-Gatherers'. *Tourism Geographies*, 1(2), 142–163.
Weeden, C., & Boluk, C. (2014). *Managing Ethical Consumption in Tourism*. New York: Routledge.
Wilson, J. (2011). Australian Prison Tourism: A Question of Narrative Integrity. *History Compass*, 9(8), 562–571.
Žižek, S. (2008). *Violence: Six Sideways Reflections*. London: Verso.

CHAPTER 6

Israel State, Genocide and Thana-Capitalism

INTRODUCTION

In a recently published book, *The Rise of Thana Capitalism and Tourism*, we used the term "Thana-capitalism" to signal a new stage of capitalism which replaced the paradigms of Ulrich Beck regarding the risk society by a new morbid consumption (Korstanje, 2016). The post-modern sociologists imagined a society where risk mediated between the lay citizens and the democratic institutions (Beck, 1992, 2009; Giddens, 1999; Luhmann, 2017). The 2000s brought substantial changes that molded a new geopolitical cosmology for the US and Europe. The attacks to the World Trade Center marked the end of an era, where a much wider process of securitization prevailed. As David Altheide (2017) observes, the culture of fear, which was already existent in the US, amplified the symbolic effects of terrorism to the extent some apocalyptic discourses were energized a new cultural industry disposed to make from horror and the end of the world a source of entertainment. The 9/11 inaugurated a new momentum of West where the media and terrorism formed a difficult symbiosis. The present chapter holds the polemic thesis that after 9/11 emerged a thana-topic culture which inverts the allegory or genuine lessons of Holocaust toward a new form of victimization.

Let's explain to the readers that the term "genocide" was originally coined by Raphael Lemkin just after the crimes committed against civilians in Nazi Germany. Although the protocols and derived legislation were designed to protect the integrity of weakest groups, it is equally true that, paradoxically, genocides—in the twentieth centuries—were perpetrated by the same nation-state that hypothetically should avoid them. Michael Ignatieff (2003) justifies torture as the lesser evil that helps the government to struggle against terrorism, and argues that Lemkin's concerns were not only the number of casualties the WWII left but also the dichotomy of nation-states, which under some conditions may very well assassinate their own citizens. The prerogative of Nazis was not ethically judged in the principle of innocent of the ethnicities they in cold blood exterminated but in the premise, Hitler's government was elected by the people. The notion of democracy, in this vein, was manipulated in both sides of the Ocean, in Germany alluding to the extraordinary destiny of a super-race created to dominate and administer the world and in the US where some minorities (like blacks) were systematically excluded (Mayer, 2013). The fact is that the concept of freedom is very hard to grasp and define. In the name of liberty or humanitarian aid, governments commit serious human right violations. Although the doctrine of the lesser evil is not convincing, it lays the foundations for a new debate with respect to the nature of evil. As Hannah Arendt (1963) puts it, the banality of evil suggests two important aspects of ethics. Firstly, evil acts—beyond any interpretation—may be the result of a bureaucratic system or a standardized form of behavior. Secondly, Nazism imposed an unethical instrumentalization of death in the name of the progress (Levinas, 1988). Nazis were insensitive to the Other's pain, while they operated within the horizons of creative destruction. Based on biological connotations, Nazis imagined a world without undesired races where the natural order ensured the survival of the strongest. Though Hitler lost the war, he won the ideological battle and the ideology sooner implanted in the US—in a mitigated form that specialists know as social Darwinism (Skoll, 2012). This chapter discusses in depth one of the paradoxes of history observed by Karl Marx where the oppressed people, once they break the ties with oppressors, are doomed to repeat the same history. Marx found an answer in the material forces of history (the Hegelian dialectics) which not only shaped but also determined the struggle of classes—as the axis of history. The revolution did the correct thing ousting the ruling elite, while paradoxically places the proletariat in the same dilemma. Sooner or later, the oppressed class does the same as their ancestors, inspiring the revolt of other classes (Marx, 2008).

The Anthropology of Holocaust

Academicians of all stripes have philosophically discussed the Holocaust as a real tragedy that involved not only an ethnic minority but also marked a turning point in the fields of human rights (Bloxham, 2001; Feierstein, 2014; Friedrichs, 2000; Lipstadt, 1993; Nussbaum, 2001; Skoll, 2012). This tragic event marked the agenda of sociology and philosophy to date. Primo Levi debates around the "drowned and the saved", while Arendt bespeaks of the banality of evil. The present section deals with the different points and argumentations revolving around the genocide and the cultural effects of the Holocaust in the Western civilization. As stated, Levi originally describes the memories of extreme experiences which need re-memoration to alleviate the trauma. Toying with the term "good faith", Levi writes brilliantly that oppressors cover their real interests—lying—to save themselves, while victims believe they suffer for reaching some divine goal. At the bottom, both are part of the same game. The process of communication plays a leading role in the struggle to survive (Levi, 2017). Richard Bernstein (2002) theorizes on the nature of the radical evil as the precondition toward modernity. Auschwitz and the violations of human dignities in the concentrations camps not only are part of human nature but will probably repeat in the future. For Bernstein, the idea of a "radical evil" coincides with the needs of knowing if the same trauma—we try to forget—can take room again. Although Westerners are certainly educated to think they are part of the selected people, a privileged group that makes what should be done, they paradoxically formatted to embrace evil. If Hegel was right in thinking the subjects are prone to adopt countless moral discourses, which vary on culture and time, history exhibits a fertile ground to gain further understanding on the complexity of genocides (Bernstein, 2002). In consonance with Bernstein, Lang Berel (1999) calls the attention to the idolatry of human rights, demonizing torturers and perpetrators of crimes against humanity. He holds the polemic thesis far from helping in making justice; these types of activists usher society into a climate of trivialization where demons escape to fair trials. As a result of this, the allegories around the Holocaust can be politically tergiversated—in witch-hunts—to protect the interests of current officialdom. When this happens, the probabilities that there are repetitions of the same event are higher, Lang clarifies.

One of the authoritative voices that focused his attention on the theme of the Holocaust was Zygmunt Bauman. From his viewpoint, the

Holocaust was something else than an event happened to Jews, but it is the direct results of the expansion of modernity in Europe. The question of whether bureaucracy allowed the Nazis to domesticate the critical thinking, as Bauman adheres, was not related to the authoritarian character some colleagues hold. Capitalism as a cultural project encompassed a logic of instrumentalization of the Other's pain, which was capitalized by Nazism. In these lines, Bauman offers a new diagnosis on the Holocaust as the emergence of the rational spirit, which derived in the needs of efficiency—no matter the costs—and scientific mentality. Nazism articulated a climate of terror and dehumanization against some ethnicities because the rise of instrumentality was adjoined to the lack of criticism and the rise of a narcissist character that revamped the already established legal jurisprudence by Nuremberg Laws (Bauman, 1988, 1989). In this respect, Slavoj Žižek contends that cruelty and violence are human conditions, but he asks why the death of some resonates further than the death of others. Žižek anticipates the political interpretations of mass death alternating the ethical borders as an ideological instrument of discipline. Today's governments appeal to the figure of the Holocaust—and others victimizations—to pose policies that otherwise would be globally rejected. Though the state of Israel claims victimization on the basis of the effects of WWII, less is discussed about the abuse of human rights of Palestine population, or even the limitations of American law to legislate into Guantanamo Bay (Žižek, 2008). The spectacle orchestrated around the death destroys any understanding or inference about the reasons behind the tragic events.

Modernity and the Consumption of Death

The French historian Phillipe Aries offers a more than interesting thesis with respect to our rites and attitudes before dying. He offers the following axiom: while in Middle Age people faced serious famines and intestine wars that decimated the entire population, the social imaginary was familiar with death. While the introduction of technology and medical health advances expanded the life expectancy, the act of dying turned in an unpopular condition only reserved to pariahs and people living in indigent conditions. The paradox lies in the fact that if we live longer, death becomes wilder, a real undomesticated best, which places our civilizatory project in jeopardy (Ariès, 1975). The figure of death was an object of study for many sociologists and philosophers over the recent decades (Barro & Ursua, 2008; Baudrillard, 1995; Chan, 2003; Klein, 2007;

Korstanje, 2014, 2016, 2018a; Stone & Sharpley, 2008). In her book *Traumascapes: The Power of Fate of Places Transformed by Tragedy*, Tumarkin (2013) coins the term, "traumascape" to symbolize those places destroyed by disasters, genocides or mass-death, which not only were economically recycled but culturally transformed in a site of attraction for visitors and tourists. As discussed in the previous chapter, dark tourism and the visit of macabre sites have situated as a new emergent segment that captivated the attention of countless fieldworkers. As Stone puts it, dark tourism denotes the idea that visitors are in quest of answers about their own lives, while theorizing and inspecting through the Other's death (Stone, 2011a, 2011b, 2012; Stone & Sharpley, 2008).

Our position is that the dark consumption covers an ideological message, which reflects the symbolic touchstone of modern capitalism. The encounter between the imperial North and the global South seems not to be easier, as the exegetes of neoliberalism thought. Tzanelli (2016) reminds that there is a utopia of guilty landscape that characterizes the consumption of peripheral tourist destinations which are often vulnerable in view of their lack of material resources to mitigate disasters or start with economic programs of economic relief. At a closer look, the capitalist world offers the cosmography for riches and another for poor nations. Thana-tourism and slum tourism follow the same dynamic, serving as a mechanism to interpret the Other's pain through the lens of consumption. The privilege aura of travelers, who come from the First World contrasts with the lives of favelados, known as the dwellers of slums in Brazil, or even those whose households have been fully destroyed by a hurricane. Capitalism never corrects the glitches that led to disasters but commoditize its effects to be gazed by First World citizens. The same nations, which centuries ago, delivered their armies to oppress the natives—in the days of colonial rule—today send tourists toward the same peripheral colonies. In this case, the violence and dispossession set the pace to the needs of captivating the Others' suffering, which suggests visual dispossession. In that way, these visitors not only reinforces their supremacy as citizens pertaining to the more "democratic cultures" but also as rational white agents of civilization (Tzanelli, 2017).

It is tempting to say Auschwitz tries to reconcile two opposing worlds. On one hand, we have the world of those who evoke the mourning of a real death or torture. On the other, tourists visit the concentration camps in quest of something new or simply to know what happened there. Can we reconcile the effects of genocide—as previously debated—with the logic of depersonalized consumption?

What Is Thana-Capitalism?

In our book the *Rise of Thana Capitalism and Tourism*, we laid the foundations of a new theory to expand the current understanding of capitalism. The risk society, as it was imagined by Beck, was no longer alive. Rather, we started the process toward a new phase where—after 9/11—society succumbed to the macabre game. Based on Baudrillard's legacy and his theory of the spectacle of disaster, we outlined a fresh conceptual model to explain the consumption of Others' death was not limited to tourism, but it was found everywhere, through the media entertainment. Basically, the society in Thana-capitalism rested on two significant tenets: (a) the needs of being different, special or outstanding—narcissism—and (b) the quest for security.

To some extent, Nazis were defeated but their most significant allegorical message was internalized in the heart of Puritan America (we are born to dominate the world). The social Darwinism showed the needs of a radical competence to mark off the elements that are to be preserved or discarded. Nazis massacred a lot of innocent people but it is important to see what they did in secrecy. In Thana-capitalism, the victims are mediatically portrayed to keep the workforce under control. As Aries noted, we live in a less conflictive world, but it paved the ways for the rise of an unsaid panic when the death arises.

The epicenter of Thana-capitalism comes from the attacks to World Trade Center in charge of Al-Qaeda; an event occurred on September 11, 2001. This shocking blow resonated as a turning point where Islam radicalism showed not only the weaknesses of West but also how the means of transport, which were the badge of US, were employed as mortal weapons directed toward civil targets. Educated and trained in the best Western universities, jihadist showed the dark side of the society of mass consumption. Many of the steps followed by Al-Qaeda were emulated from a management guidebook. This made-man disaster showed the proud USA that regardless of strength, power or levels of development of the country, terrorism will be present in all central nations. From this moment on, nobody will feel safe anytime and in any place. As Catholic Church in Lisbon's Quake through 1755, the rational basis of risk experts or risk-related analysis was placed under the critical lens of scrutiny. Beyond what radical conservative in Bush's administration precluded, this event initiated a new age where the concept of security and prevention start to dilute.

The classic scale production which characterized the old industrialism mutated toward chaotic, complex and unpredictable realms, accelerated after the stock and market crisis in 2008, where the atomized demands become the competence of all against all (in the Hobbesian terms). The Darwinist allegory of the survival of the strongest can be found at the main culture value of Thana-capitalism in a way that is captivated by cultural entertainment industries and cinema. Films such as *Hunger Games* portray an apocalyptic future where the elite govern with iron rule different colonies. A wealthy capitol which is geographically situated in Rocky Mountain serves as an exemplary center, a hot-spot of consumption and hedonism, where the spectacle prevails. The oppressed colonies are rushed to send their warriors who will struggle with others to death, in a bloody game that keeps people excited. Although all participants work hard to enhance their skills, only one will reach the glory. The same can be observed in reality TV show such as *Big Brother*, where participants neglect the probabilities to fail simply because they over-valorize their own strongholds. This exactly seems to be what engages citizens to compete with others to survive, to show "they are worthy of surviving". In sum, the sentiment of exceptionality triggered by these types of ideological spectacles disorganizes social trust. Capitalism signals to the constructions of allegories containing death prompting a radical rupture of self with others. Whenever we see ourselves as special, we put others of different condition asunder. In a context of turbulence, the imposition of these discourses is conducive to the weakening of social fabric. Thematizing disasters by dark tourism consumption patterns entails higher costs in the form of disaster repeating in the near future. The political intervention in these sites covers the real reasons behind the event, which are radically altered to protect the interests of the status quo. The political and economic powers erect monuments to symbolize sudden mass death or trauma spaces so that society reminds a lesson, which allegory contains a biased or galvanized explanation of what happened. George H. Mead realized, some time back, that audiences questioned the news relating to murders, while they are obsessed with reading them. He concludes that the self is configured through its interaction with others. This social dialectic introduces anticipation and interpretation as the two pillars of the communicative process. The self feels happiness through the Other's suffering—a rite necessary to avoid or think about one's own potential pain. Starting from the premise that the self is morally obliged to assist the other to reinforce a sentiment of superiority, avoidance preserves the ethical base of social relationships (Mead, 1967).

Mead's reflections could be applied to the act of visiting dark tourism shrines. To understand this, we can revert to the myth of Noah and its pivotal role in the salvation of the world in Christianity. The legend tells us that God, annoyed by the corruption of human beings, mandated to Noah to construct an ark. Noah's divine mission consisted of gathering and adding a pair per species to his ark so as to achieve the preservation of natural life. The world was destroyed by the great flood, but life diversity survived. At first glance, the myth's moral message is based on the importance of nature and the problem of sin and corruption. But when examined more carefully, the myth poses the dilemma of competition: at any "tournament" or game, there can be only one winner. In the archetypical Christian myth, Noah and the selected species stand as the only witnesses of everything and everyone else's death. We argue that the curiosity and fascination for death come from this founding myth, which is replicated in plays to date, stating that only one can be crowned the winner. Even, the *Big Brother* show, which was widely studied by sociologists and researchers of visual technology, rests on this principle. Only a few are the selected ones to live forever on the screen, as is the case in religious myths such as those of Protestantism and Catholicism (both based on doctrines of salvation and understandings of death).

The consumption of disasters as well as mass death reminds us how special we are, or affirms our ego into the privileged status of those who escaped to death. Paradoxically, this revives a morbid form of consumption which plays a leading role in the configuration of new capitalism where death is the main commodity to exchange. As explained in the preceding section, the plot of the novel *Hunger Games* is illustrative in how Thana-capitalism works. The capitol exploits 13 districts at the hands of the cruel President Snow. Each district should deliver representatives who fight to the death in a mediatized spectacle. While each participant over-valorizes its own probabilities to defeat the others, Snow and the capitol turns stronger. In a Darwinist scenario, where only the strongest survive, death is the end of everything for the thousands of oppressed peoples. The consumption of Others' suffering through the tube offers the wider spectatorship a reason to belong to the society, while stimulates the competence as an ideological discourse of what is being "a good citizen". Winners do not help others because the winner takes all! This logic feeds a climate of Darwinism where the Others' death situates as a criterion of stimulation, entertainment, if not happiness. To put this bluntly, gazing these types of sites of mass destruction seems to be something else than a simple fashion, but a change of age. It exhibits only the peak of the iceberg of a much deep-seated issue, which merits to be discussed in the next approaches.

In sum, Thana-capitalism draws a new ideology of competence which commoditizes death for First World cultures to reinforce their auratic sentiment of supremacy over the periphery. For that reason, we conclude that the Holocaust and Thana-capitalism are inextricably intertwined. The current chapter focuses on the theatralization of disasters as the main commodity Thana-capitalism exchanges. The discussion around the crimes against mankind perpetrated by Nazis in the clandestine concentration camps opened the doors toward new insights with respect to the roots of Thana-capitalism. Nazis violated the human rights by secreting their crimes, in a moment of the world where millions have certainly died. Today's philosophers are shocked to see how Auschwitz-Birkenau, which was the sanctuary of the horrors of WWII, sets the pace to a new allegory, intended to entertain thousands of tourists, many of them unfamiliar with these events. As a highly demanded tourist destination, Auschwitz evinces the change of new post-modern ethics that commoditizes the Other's loss as a criterion of entertainment. The example of terrorism shows one of the paradoxes of Thana-capitalism simply because of media coverage and disseminates the cruelties of attacks in order to gain further subscribers and investors, while terrorism finds a fertile ground to enter at the homes of a wider audience. The audience seems to be devastated by the breaking news that broadcast the moment people blast or is being decapitated, but at the same time, they are unable to stop watching! Lastly, Thana-capitalism has successfully disorganized the social ties through the imposition of allegories, dark landscapes and the spectacle of disaster in order for the status quo to keep untouched.

THE STATE OF ISRAEL AND THE POWER OF PROPHECY

A great portion of the Christian myths, such as the exile of Adam from Eden and the Noah Ark, comes from Jewish tradition. In fact, one must confess the history and lore of Israel is not only conniving but also fascinating. In regards to the prophecy, Professor of Old Testament Studies at the University of Notre Dame, Joseph Blenkinsopp (1983) clarifies, its function is given for the protection of community and in denouncing the arbitrariness of officials and kings. The prophecies can be divided into critical and optimistic ones. For this reason, one might infer that prophecies not only worked in an earlier prophetic cult but also varied and resonated in different ways in the history of Judaism. Most probably, the countless interpretation prophecies often take, as well as the political adjustment of their content to the established political power makes very difficult to examine them as an all-pervading conceptual corpus. What Blenkinsopp found very conniving appears to be that from Elijah to Jonah, prophecies are historically placed in the tradition of Near East when Jews were enslaved by Babylonian and Assyrians. Following this, the different political turmoil and disasters happened through the sixth century BCE, shaped an emerging wave associated to the needs of exile. The figure of prophecy played a leading position by giving the community certainness in an ever-changing environment.

> The political disasters of the early sixth century BCE would have provided a further stimulus to the preservation of prophetic sayings, and it would be surprising it efforts had not been made during the exile to put together a selective corpus of such saying. The editorial history of the prophetic books, especially Jeremiah, suggests that Deuteronomic scribes played a significant role at that time. The process of editing and expanding this material continued into the Second Temple period. (Blenkinsopp, 1983: 23)

Starting from the premise that Moses receives Torah from God at Sinai and pass the legacy to his son and the son of his son, the function of the prophet is the creation of a bridge between God (as the source of this primordial revelation) and the rabbinic authorities. As the custodians of the Torah, prophets and their visions were strictly subordinated to this mentioned sacred text; however, in the time, they became critical voices of the political order, even to the brink of causing political instability. In the nineteenth century, some theologians erroneously confirmed that prophets were the forerunners of Christ reminding the conflict of this former with the established rabbinic power. Nevertheless, the modern scholars acknowledged that:

> Modern critical scholarship, which did not derive its mandate from ecclesiastical authority, studied the prophets independently of such a traditional loci communes. Applying literary criticism to the task of identifying the actual words of the prophets, their authentic message as distinct from secondary editorial accretions, it claimed to find in them a unique class of religious individualists with a message focused on the present rather than the distant past or the distant future. (p. 27)

Although romantic scholars stressed the figure of some prophets over others, as Blenkinsopp asserts, the prophets formed a closed group where the mysterious experience between them and God was the essence. Hence, the prophecy is, above all, an unchangeable experience. Blenkinsopp contends that the compilation and edition of these oral sayings were accompanied by the political instability Jews continuously faced in their history.

In this respect, Barry Horner (2001) reminds that the consistencies of prophecies in Israel history were tergiversated by Christianity, at least biblically speaking. The tension between Christ and the anti-Christ was conditioned by the political map of Middle Age. He is convinced that there is a "wrong perception of Israel" (p. xix) by Christians which created undesirable effects in the history of the two religions. The Reform, in the later centuries, developed a negative and hostile view against Judaism, which merits to be reconsidered. Unlike Rabkin and Arendt, Horner struggles for a *national Israel* as the suitable interpretation of rabbinical texts. From the establishment of Israel State in 1948, an act which was accompanied by the reclamation of Jerusalem in 1967, there was a direct conflict between Palestine and Zionists. Though Christianity shares many commonalities with Zionism, it is equally true that a dark future was augured for Israel. The scatological tradition in Christianity speaks that the national Israel, dotted of a specific territory, would be the sign of the end of the world. Based on the Augustinian legacy, anti-Zionist Christians endorsed a great sympathy for Palestine State considering Israel an illegitimate occupier.

Over the recent years, some scholars have exerted a radical criticism on the politics of Israel State regarding Palestine. Some human rights violations, associated to an endless conflict, inspired the ink of radical philosophers as Slavoj Žižek (2011) or Andre Glucksmann (2005). While the former discussed the sentiment of victimization that evokes the Israel government each time the world questions its actions, the latter calls for the attention to the cynic role of France and others European countries that never promoted the creation of a state in Europe. Centuries of anti-Semitism and xenophobic expressions make the task to analyze the issue very hard. Rather, this section echoes the original concerns of Hannah Arendt (see Young-Bruehl, 1982) and Yakov Rabkin (2006) in conceiving the Jewish culture as a nomad and itinerant project which would never crystalize in a nation-state. Judaism becomes a threat from within at the time it converted in a territorial state. Rabkin argues convincingly that Zionism claims itself as the only opportunity to establish a state that represents all Jews and Jewishness. However, far from representing all interests, the Israel State was historically an object of critiques even to the Jewish community. At a first glimpse, Zionism not only emulated the Leninism doctrine in Russia but looked to control the territory calling for all Jews in the world. This was not possible without a radical transformation of Jewish identity similar to other European nations. Zionism centers on those European Jews who have been historically relegated and unassimilated in the old continent, dangling that they move by a much deeper feeling of frustration (even ignoring the sacred mandate emanated from Torah). As a secular movement, Zionism overlooks the rabbis. Zionism, for Rabkin, was an invention of the European intelligentsia which confronted with the real spirit of Judaism (based on a multicultural tradition). During years, Judaism was systematically repressed, silenced and left to the dust of oblivion, while Zionism occupied central roles in the Israel State. Rabkin explores the dichotomies of Israel State and the human right violations in Palestine. Basically, as a European project, this state resulted from a nationalist discourse opposed to the cosmopolitanism that marked Judaism from its inception. The process of secularization instilled by Zionism changes the cosmologies of Jews. The classic Jew embraces Jewishness by his acts and not as a form of identity. From the nineteenth century, European Jews, as well as the Zionist movement, situate as a modernizing force in opposition to the past and tradition, whereas paradoxically reframe a fabricated past to idealize the Jewishness as a millennial dream. The messianic hope sets the pace to a recalcitrant nationalism, where rabies plays a bit-role. The divine sphere of God, where Jews are judged by their sins (e.g., the Diaspora is a result of the sin of Jews not a disposition of Roman Empire) is replaced by the reason of state. The concept of heritage is imported to start a reformist process that calls to Jewishness as its symbolic touchstone. The fact is that the love by the soil, which is inspired by Zionism, keeps an ideological nature, Rabkin concludes.

In addition, those thinkers and scholars (like Arendt) who confronted Zionism were energetically rejected or at the best criticized. In consonance with this, Norman Finkelstein (2000) employs the word "the holocaust industry" to refer to the cultural consumption, as well as the propagandistic indoctrination articulated around the figure of the Holocaust. Organized revolving around a unilateral dogma, Finkelstein holds, the archetype of the Holocaust has become a "cult" which reclaims a historical uniqueness that never existed. This means that genocides are not the monopoly of Jewish people or part of any other ethnicity; it is a condition of exploitation and eradication administered by one group over the others. As Finkelstein puts it:

> All holocaust writers agree that the Holocaust is unique, but few, if any, agree why. Each time an argument for Holocaust uniqueness is empirically refuted, a new argument is adduced in its instead ... uniqueness is a given in the Holocaust framework; proving it is the appointed task, and disapproving it is equivalent to Holocaust denial. Perhaps the problem lies with the premise, not the proof. Even if the Holocaust were unique, what difference would it make? How would it change our understanding of the Holocaust were not the first but the fourth or fifth in a line of comparable catastrophes. (Finkelstein, 2000: 121)

The mystification of Holocaust negates history running a serious risk to repeat the tragic events. This situates the survivors in an exemplary position or at the best in a mirror where others similarly tortured victims cannot be reflected. The Holocaust seems to be unique and special because Jews are special and outstanding. Unless otherwise resolved, this thought leads toward a strange sentiment of exceptionality that nourishes ethnocentrism (Finkelstein, 2000). As discussed, the roots of Thana-capitalism, which was encapsulated in the Puritan cosmology, is not pretty different to this dogma. We need the pain to mark our destiny on the earth, to prove ourselves we, after all, are stronger, smarter, tougher than the rest (Korstanje, 2016). Finkelstein's text is useful to our argumentation because of two main reasons. First and most important, the arbitrariness of Israel State—for example, in the Lebanon invasion in 1982—is widely justified through an emptied meaning of Holocaust comparing, so to speak, the Arabs with Nazis. This moot point is assessed by Slavoj Žižek and other colleagues. Secondly, the process of victimization which legitimates oppressive policies enables anti-Semitic reactions in the opposite side (Finkelstein, 2000). In her book *Denying Holocaust*, Deborah Lipstadt (1993) clarifies that beyond the deniers, who question the official history of the Holocaust in the days of Nazi Germany, there lies new subtle racism aimed at silencing the position of Jews in the history.

As this argument is given, Didier Fassin (2009) speaks of the Empire of Trauma as the major signifier of our times. The work is originally destined to describe how the subject infers different political uses of its traumatic re-memoration. The twilight of 9/11 not only shows the exposure to traumatic events in the media or raises the probability to suffer mental diseases as depression or drug addiction, but there is also a vicarious trauma that affects the public audiences and television viewers when they are subject to apocalyptic landscapes. Fassin acknowledges that events like 9/11 leave scars, which affect the society collectively, as well as endurable effects in the individual behavior. Rarely evoked some decades ago outside the hegemony of psychology and medicine, nowadays trauma associates to the status of victimhood. As he brilliantly observes:

> Our goal is rather to understand how we have moved from a realm in which the symptoms of the wounded soldier or the injured worker were deemed of doubtful legitimacy to one in which their suffering, no longer contested, testifies to an experience that excites sympathy and merits compensation. (Fassin, 2009: 5)

Fassin goes on to say that this happens not only because the psychiatry exerted a notable influence over other disciplines extending the notion of trauma to those bodies disciplined by the violence but, in fact, the society developed some sentiments of empathy over the victims. It is remarkable that when the story is somehow distorted, repressed or veiled, manifest violence suddenly emerges in the body or the mind. The genealogy of trauma involves a dual nature. On one hand, the concept was historically monopolized by the medical sciences (psychiatry and psychology). On the other, there is a second interpretation which associates to the social construction revolving around the tragic event. Although the first point is widely examined in medicine, the second is very hard to investigate. Fassin intends to fulfill a gap in this field. His thesis seems to be that the idea of trauma helps the subject to have further support in the society, while as a moralizing force it carefully selects who is worthy of being touched and forgiven. In this way, each biography, which is historically and culturally determined, is interrogated and replaced by a universal archetype of trauma. In the Empire of Trauma, the pain is often manipulated and commoditized to create a political order (Fassin, 2009).

WHAT IS THE ROLE OF PROPHECY IN THIS PROCESS?

In a recent book, entitled *Essays in Political Anthropology*, we critically question the nature of the liberal state and the paradoxes it represents. Based on two seminal myths, the fall of Troy (Greek literature) and *Star Wars* (2005) (George Lukas' film), we offer an all-encompassing model to understand how the prophecy, which means a closed future, transforms good desires into pure evil. Agamemnon is a King, who taking advantage of the lack of hospitality of Paris, invades Troy (accompanied by Aquila and other legendary warriors). After years of war, Agamemnon is trapped against the wall and the deep blue sea and opts to sacrifice his daughter to Poseidon. Finally, and once the sacrifice is consummated, he enters Troy though he released the tragedy. Assassinated by his wife, the myth tells that governor does what should be done—no matter the ethical consequences. However, at a closer look, another interpretation is needed. Agamemnon is not interpellated by the future. Though he asked the advices of the Gods, he always keeps the final decision to himself. His future, so to speak, is open to his individuality. Rather, the young Skywalker succumbs to the dark side of the force pressed by a closed future. He dreams how his lovely Padme dies during childbirth. Obsessed by the looming future, Skywalker is seduced by Darth Sidious (Senator Palpatine) who affirms to have the solution for Padme. Skywalker falls to the dark side as a product of his love and the desperation because the future cannot be altered.

For those who have not viewed the film let's clarify how the plot evolves. Skywalker faced a traumatic experience—his mother's death. This not only marked him forever but also accompanied his darker dreams. In one of these episodes, he dreams his love Padme dies at the time of giving birth to two babies (Luke and Leia). In contrast to Agamemnon, Skywalker loves Padme, but what is more important, her death is a premonitory dream. Padme's death is predestined and Skywalker's fears aggravate once he receives the news she is pregnant with twins. Skywalker knows the grim future, while he is hand-tied to change it. In one of his consults, Yoda alerts him on the risk when future overwrites the present. Unfortunately, his powers are not enough to reverse the roots of the tragedy as well as his decline of autonomy with respect to the dark side. Obsessed by his dreams, the youth Jedi looked for the source of eternal life, which is mysteriously offered by a Lord Sith, Darth Sidious. Skywalker is enmeshed in the philosophical angst of predestination, whereas a new encroachment by the Sith starts. Sidious finally promised Skywalker to give the secret of eternal life, a covered secret keeps by Jedi cult but which the Siths may amply and generously share. Lastly, Skywalker not only succumbs to the dark side but, once converted, he is also commanded by two cruel missions. He is delivered to Jedi Temple and kills any resistance there, even children! Skywalker does not conduct a sacrifice to win the city; he has killed others for salving his love (Korstanje, 2018b). The lesson—left by the two plots—

shows how the power of prophecy closes the horizons of future activating the precautionary dogma that leads a good person to embrace the evilness. For centuries, Israel was enmeshed in its own prophecy accepting suffering as a valid path toward the encounter with the Lord. As an errand culture (nomad in the desert), the belief in a promised land characterized the hopes and fears of the Jewish community. The prophecy played a crucial role in giving these desperate people a reason to continue but at the same time, it created a closed vision of future or a blind trust in God. The crimes perpetrated by Nazi Germany were foreseen but permissible inaugurating a new era that allowed the creation of an Israel State, a sacrifice that after all has a happy end. This cosmology, which is structured on the needs of suffering to be redeemed and the importance of prophecy to remind how important Jewish culture is, results in the Thana-capitalism and the quest of uniqueness. The question of whether Nazis killed millions of civilians they perpetrated their macabre plans in secrecy. Now, in Thana-capitalism, death situated as a show which serves as an ideological form of entertainment. Probably, Zionism—echoing Rabkin—adopted and replicated the sense of predestination—borrowing this tradition from Europe—in their own prophecies, but what is clear is that this is a deep-seated issue which merits to be investigated in other approaches.

Conclusion

Capitalism should be understood as a cultural project, besides an economic system, which is based on two preliminary aspects: social Darwinism and the doctrine of predestination enrooted in the Protestant Spirit. Two scholars have explored with brilliant mastery on both, Max Weber (2002) and Richard Hofstadter (1944). While the former signaled to capitalism as a consequence of Protestant Reform that divorced from Catholic Church, the latter envisaged that social Darwinism was the key factor to grant the competence necessary for market expansion. Social Darwinism was a theory coined by Sir Francis Galton, whose interests were oriented to adapt the concept "evolution of species" as it has been delineated by Charles Darwin into the social world. However, Galton not only misjudged Darwin's advances in the fields of biology but also confused "the survival of the fittest" with "the survival of strongest". In contrast to Darwin, social Darwinism observed that natural selections can be applied to social scaffolding. In a way, some species struggle with the environment to survive; humans struggle with others to reach success. In this token, the

Anglo race was placed on the top of the social pyramid as the most evolutionary ethnicity with respect to other minorities. At the same time, this doctrine paved the way not only for racist ideas in America that shaped capitalism, but also Nazism in Europe. In parallel, as Hofstadter puts it, the idea of an exemplary race or dreams of uphill city contributed to a discourse of superiority of Anglo-Saxons over other cultures, which sooner or later encouraged "the war of all against all"; social Darwinism works because rank-and-file workers struggle with other workers by a job or better opportunities. While capital-owners monopolize their power into few hands, workforce is atomized to avoid the unionization. Those who have not developed adaptive skills to survive are considered "the weak". After all, capitalism always grants the survival of the strongest, the best agent. In the fields of religion, Weber anticipated a similar landscape. Capitalism was the result of Protestant logic of "predestination", which means that the soul's salvation was pre-determined by Gods in the life-book. Only few will be gathered by the Lord in the bottom-days. For wayward Protestants, the world not only is a dangerous place but also the platform to show one deserves the salvation. The force of labor seems to be the sign marking the boundaries between doomed and salved souls. This is the main cultural difference between Catholics and Protestants. However, some insights reveal that the power of predestination can be traced to the Judeo-Christian background.

REFERENCES

Altheide, D. (2017). *The Politics of Terrorism*. New York: Rowman & Littlefield.
Arendt, H. (1963). *Eichmann in Jerusalem*. New York: Penguin.
Ariès, P. (1975). *Western Attitudes Toward Death: From the Middle Ages to the Present* (Vol. 3). Baltimore: Johns Hopkins University Press.
Barro, R. J., & Ursua, J. F. (2008). Consumption Disasters in the Twentieth Century. *The American Economic Review, 98*(2), 58–63.
Baudrillard, J. (1995). *The Gulf War Did Not Take Place*. Bloomington: Indiana University Press.
Bauman, Z. (1988). Sociology After the Holocaust. *British Journal of Sociology, 39*(4), 469–497.
Bauman, Z. (1989). *Modernity and the Holocaust*. Ithaca: Cornell University Press.
Beck, U. (1992). *Risk Society: Towards a New Modernity* (Vol. 17). London: Sage.
Beck, U. (2009). *World at Risk*. Cambridge: Polity Press.
Bernstein, R. J. (2002). *Radical Evil: A Philosophical Interrogation*. Cambridge: Blackwell Publishers.

Blenkinsopp, J. (1983). *A History of Prophecy in Israel*. Philadelphia: The Westminster Press.

Bloxham, D. (2001). *Genocide on Trial: War Crimes Trials and the Formation of Holocaust History and Memory*. Oxford: Oxford University Press.

Chan, E. (2003). War and Images: 9/11/01, Susan Sontag, Jean Baudrillard, and Paul Virilio. *Film International*, 5, 128–147.

Fassin, D. (2009). *The Empire of Trauma: An Inquiry into the Condition of Victimhood*. Princeton: Princeton University Press.

Feierstein, D. (2014). *Genocide as Social Practice: Reorganizing Society Under Nazis and Argentina's Military Juntas*. New Brunswick: Rutgers University Press.

Finkelstein, N. (2000). The Holocaust Industry. *Index on Censorship*, 29(2), 120–129.

Friedrichs, D. O. (2000). The Crime of the Century? The Case for the Holocaust. *Crime, Law and Social Change*, 34(1), 21–41.

Giddens, A. (1999). Risk and Responsibility. *The Modern Law Review*, 62(1), 1–10.

Glucksmann, A. (2005). *El Discurso del Odio (The Discourse of Hate)*. Madrid: Taurus.

Hofstadter, R. (1944). *Social Darwinism in American Thought*. Boston: Beacon Press.

Horner, B. E. (2001). *Future Israel: Why Christian Anti-Judaism Must Be Challenged*. Nashville: B&H Academic.

Ignatieff, M. (2003). *Human Rights as Politics and Idolatry*. Princeton: Princeton University Press.

Klein, N. (2007). *The Shock Doctrine: The Rise of Disaster Capitalism*. New York: Macmillan.

Korstanje, M. E. (2014). Chile Helps Chile: Exploring the Effects of Earthquake Chile 2010. *International Journal of Disaster Resilience in the Built Environment*, 5(4), 380–390.

Korstanje, M. E. (2016). *The Rise of Thana Capitalism and Tourism*. Abingdon: Routledge.

Korstanje, M. E. (2018a). A Paradoxical World and the Role of Technology in Thana-Capitalism. In M. E. Korstanje (Ed.), *Encyclopedia of Information Science and Technology* (4th ed., pp. 4761–4773). Hershey: IGI Global.

Korstanje, M. E. (2018b). *Essays in Political Anthropology: Reviewing the Essence of Capitalism*. New York: Nova Science Publications.

Lang, B. (1999). *The Future of the Holocaust: Between History and Memory*. Ithaca: Cornell University Press.

Levi, P. (2017). *The Drowned and the Saved*. New York: Simon & Schuster.

Levinas, E. (1988). Useless Suffering. In R. Bernasconi & D. Wood (Eds.), *The Provocation of Levinas: Rethinking the Other* (pp. 156–165). London: Routledge.

Lipstadt, D. (1993). *Denying the Holocaust, the Growing Assault of Truth and Memory.* New York: Free Press.
Luhmann, N. (2017). *Trust and Power.* New York: John Wiley & Sons.
Marx, K. (2008). *The 18th Brumaire of Louis Bonaparte.* New York: Wildside Press LLC.
Mayer, M. (2013). *They Thought They Were Free: The Germans, 1933–45.* Chicago: University of Chicago Press.
Mead, G. H. (1967). *Mind, Self, and Society: From the Standpoint of a Social Behaviorist (Works of George Herbert Mead, Vol. 1).* Chicago: Chicago University Press.
Nussbaum, M. C. (2001). *The Fragility of Goodness: Luck and Ethics in Greek Tragedy and Philosophy.* Cambridge: Cambridge University Press.
Rabkin, Y. M. (2006). *A Threat from Within: A History of Jewish Opposition to Zionism.* New York: Zed Books.
Skoll, G. R. (2012). Ethnography and Psychoanalysis. *Human & Social Studies. Research and Practice, 1*(1), 29–50.
Star Wars. (2005). Episode III (The Revenge of the Sith). George Lukas (Dir). 140 Minutes. English. US, Lucasfilm Ltd.
Stone, P., & Sharpley, R. (2008). Consuming Dark Tourism: A Thanatological Perspective. *Annals of Tourism Research, 35*(2), 574–595.
Stone, P. R. (2011a). Dark Tourism: Towards a New Post-Disciplinary Research Agenda. *International Journal of Tourism Anthropology, 1*(3–4), 318–332.
Stone, P. R. (2011b). Dark Tourism and the Cadaveric Carnival: Mediating Life and Death Narratives at Gunther von Hagens' Body Worlds. *Current Issues in Tourism, 14*(7), 685–701.
Stone, P. R. (2012). Dark Tourism and Significant Other Death: Towards a Model of Mortality Mediation. *Annals of Tourism Research, 39*(3), 1565–1587.
Tumarkin, M. (2013). *Traumascapes: The Power and Fate of Places Transformed by Tragedy.* Carlton, VA: Melbourne University Publishing.
Tzanelli, R. (2016). *Thana Tourism and Cinematic Representation of Risk.* Abingdon: Routledge.
Tzanelli, R. (2017). *Mega-Events as Economies of the Imagination: Creating Atmospheres for Rio 2016 and Tokyo 2020.* Abingdon: Routledge.
Weber, M. (2002). *The Protestant Ethic and the Spirit of Capitalism: And Other Writings.* New York, NY: Penguin Books.
Young-Bruehl, E. (1982). *Hannah Arendt: For Love of the World.* New Haven: Yale University Press.
Žižek, S. (2008). *Violence: Six Sideways Reflections.* London: Verso.
Žižek, S. (2011). *Living in the End Times.* London: Verso.

CHAPTER 7

Disasters in the Society of Fear

INTRODUCTION

While the political power sets the national agenda with respect to many themes such as delivering troops to the battlefront, reducing taxes and ecological issues, it is no less true that lay people passively face the effects of decisions that are made by the politicians. The bad decisions made at the top of the pyramid surely engender risks, which place the ontological security of society in jeopardy (Cannon, 2008; Faulkner, 2001; Korstanje, 2010). This point leads us to think that though the contemporary society reaches the highest levels of comfort and production, lay citizens are pressed to live in vulnerable conditions. This chapter centers on a deep-seated issue, which was uncovered by the social anthropology, disasters and the cultural reactions of the ruling elite to preserve their status quo. This chapter, henceforth, defines humanitarian disasters as extreme stages of vulnerability over some groups, or ethnicities aggravated after a disaster took hit. Experts have reached consensus on the fact that humanitarian disasters frequently represent a serious hazard for the common well-being. Humanitarian crises, no less true, range from genocides or mass slaughters to ongoing states of war and acts of terrorism without mentioning the famines, ethical conflicts and epidemics. Readers understand one of the examples of humanitarian disasters is the Syrian Civil War and the refugee crisis that today places Europe in an ethical quandary (Cronin, 2015). As Russell Dynes puts it, far from being divine mandates, disasters come from concrete human decisions. The poor society, which is more devastated in post-disaster contexts, needs material and financial assistance in order for accelerating the recovery timeframe. At least, this is the idea some experts and the social imaginary hold. At a preliminary look, disasters, development and the economy seem to be inextricably intertwined.

Dynes (2002) certainly acknowledges that in the post-disaster contexts, the integrity and the responses of society depend on the synergy between public and private organizations. However, it seems equally true that some disasters are created in the name of prosperity and development (Aguirre & Quarantelli, 2008; Dynes & Quarantelli, 1975; Hannigan, 1976). Ulrich Beck was a pioneer validating the thesis that the technical breakthrough generated by modern society to make of this life a better place paved the ways for the rise of a climate of uncertainty, which paradoxically engendered new uncontemplated risks. Chernobyl sounds as a sad event that marks the beginning of a new epoch, where risks mediate between peoples and their social institutions (Beck, 1992, 1996).

As this backdrop, the existent literature overlooks how the political obsession—in generating appropriate conditions for life—may paradoxically usher society into a humanitarian crisis. It is important to discuss critically to what extent humanitarian disasters denote a much deeper background internally framed in the history of capitalism. In this respect, we confront the arguments of three senior philosophers, who from different angles, have worked on the problem of ethics, consumption and poverty: Alicia Entel, Slavoj Žižek and Kai Erikson.

The first section scrutinizes part of the literature on humanitarian disasters whereas the second deals with the idea of resilience, which is sometimes packaged as an ideological product that gives hopes to survivors and victims. Basically, what is implicit in fourth and fifth sections, the popular parlance adopts a one-sided discourse revolving around poverty. Under the logic of blaming Others, poverty serves as an ideological discourse that legitimates the ruling elite. The power of ideology is not given by what expressly it says, but what it covers. Some critical voices have pointed out that the ruling class, which often makes the decisions, usually avoids the risks it generates. Instead of facing the effects, the ruling elite mark the low classes as the real cause of the disaster. Likewise, the focus of the media is posed on the poverty and more vulnerable classes, in which case, one might imagine that humanitarian disasters take hit because poverty exists. The cure, of course, is more capitalism!

Let's remind readers that for the liberal minds and writers, capitalism should be conceived as the best of possible worlds. This creates a paradoxical situation simply because the material asymmetries generated by capitalism are never questioned. In this way, while governments devote their resources and assistance to curb poverty, further and troublesome humanitarian disasters upsurge.

Humanitarian Disasters

Doubtless, Ulrich Beck (1992) was one of the pioneers in exploring the dichotomies of capitalism as well as the effects of risk in society. The society, per his viewpoint, faces a radical shift, which can be evidenced in the family disintegration. This suggests that the economic means of production are changing to new decentralized forms that equaled all classes in the vertical sense of reflexibility. The knowledge of experts not only is confronted by lay people but also mined the trust of citizens in their political leaders. The modernity, as Beck imagined, started with the nuclear accident of Chernobyl. As a founding event, Chernobyl initiated a new logic that replaced the old industrialism not only blurring the borders of classes but also posing risk as the touchstone of society. This new spirit of capitalism reapproximated the doctrinal bases of economy introducing mass consumerism as the main ideal to follow. The orthodox literature, which was based on the society of producers, sets to a society of consumers. This begs a more than an interesting question, what is the role of fear or risk in the process?

Beck has not an accurate answer to this above-noted question, but he dangles the possibility that capitalism produces not only a set of commodities, which are globally exchanged, narrowing cultures in a hyperglobalized world but produces human wastes that mean impoverished classes relegated from the wonderland. As a result of this, the previous logic of appropriation that characterized the industrial world (or society of producers) was replaced by its antithesis, *the logic of disavowal*. Not surprisingly, mass consumption fabricates poverty and many other collateral damages, which are covered by the status quo. Beck polemically holds the thesis that capitalism operates under the auspices of a cosmology, which is legitimated by the intervention of Science and Journalism as key factors that mold popular opinion (Beck, 1992, 1996). In consonance with Beck, Bauman argues convincingly that *derivative fears*, which are a post-modern construction, resonate heavily in the social imaginary. These fears can be understood as (a) those dangers, which jeopardize the ontological security of the self, (b) those dangers which threaten the social order and the existing law and (c) the existence of the self in the world. Of course, like risk, any derivative fear not always denotes a real danger derived from a psychological feeling of vulnerability or insecurity. The ongoing climate of fear is stimulated by the status quo to indoctrinate the workforce (Bauman, 1990, 2006, 2013). Those demands emanated from the citizens that jeopardize the productive system are demonized, repelled and rejected by the ruling elite. In so doing, the fear (to a stranger or to terrorism) plays a leading role in accelerating the social fragmentation.

In this vein, the French philosopher Robert Castel agrees with Bauman that capitalism lays the foundations to a climate of mistrust and anxiety. Castel reveals an uncanny paradox, which was introduced by the late capitalism. The medieval peasant was enmeshed in an atmosphere of extreme violence, illness, and the lowest expectancy of life that offered a grim perspective. Rather, the modern citizen enjoys a set of material benefits (as never before) that expanded not only its quality but also its expectancy of life. While in Middle Age, people were subject to countless risks and dangers, they feel more secure than in modernity. The paradox is given by the fact that modernity erodes the social ties necessary for social cohesion. In this respect, the current climate of fear results from the irreversible disintegration of social order (Castel, 2006). As Castel brilliantly observed, today's mass media fabricates and packages countless risks, which never take a room in reality. The discourse revolving around the notion of risk (as an open future) comes from the needs of controlling the overabundance of contingency. With this assumption in mind, other sociologists have widely discussed the intersection of risk and contingency as Luhmann (2017) who envisaged a clear division between threats and risks. While the former signals to external threats imposed on the society, the latter are the product of a previous decision. He even says that those who made the decisions never face the risks of their decisions. To set an example, a terrorist attack is not a risk for the victims because they had any possibility to avoid the attack (Luhmann, 1993). In the opposite, Anthony Giddens understands that the risk should be contemplated beyond the individuality of the agency. At the bottom, the society of information, there is no way of living without deciding. The expansion of modernity introduced a logic of reflexibility where the produced knowledge is affordable to lay person. This opens the doors for a horizontal reflexibility that interrogates the ontological security of self (Giddens, 1991, 1999). To some extent, Giddens is influenced by the Weberian studies, which suggests that the principle of predestination posed on lay workers an easier-than-riskier position in view of the impossibility for the future to be altered. For the Protestant culture, only God knows and dictaminates a predestined fate, which can never be grasped by the human mind. This was corroborated by existentialists who alerted on the problems of modernity. They embraced enthusiastically the idea that the freedom, as it was conceived for Socratics, does not come without angst, because in the openness the self needs to decide. No matter the lines, any decision will create effects and counter-effects which constitutes the essence of humanity. Unlike animals, who move instinctively, humans decide and should bear the contingency behind their acts (Heidegger, 1962; Sartre, 1969). Sending his troop to the battle, the Commander does not know what will happen. His orders not only are obeyed, he keeps the responsibility for the security of his soldiers. His decisions are subject to the contingency that means the possibility (and the moral duties) of being the only one who may avoid the battle.

As the previous argument given, Luhmann's contributions are of paramount importance in this discussion simply because he ignites a hot debate around the political nature of risk. Like the cited case of the General who may care his troop before the battle, those who made the decision rarely face the consequences. This opens a gap between the privileged and working classes. The victims of disasters never imagined their sad fate and that the political elite are prone to blame others for their bad decisions (Luhmann, 1993).

It is widely assumed that poverty seems to be one of the maladies of our days (Moynihan, 1969; Rothstein, 2008) or in other terms a stable condition of existence that sooner or later vulnerates the human rights (Ferraz, 2008; Pogge, 2008). Some voices lamented that poverty facilitates the outbreaks of virus, illness and other social maladies which may very well place the society in jeopardy. In view of this, the communicative process, which is fostered by the mass media, plays a leading role to mitigate the effects of disasters on the hapless lower classes (Haider, Ahamed, & Leslie, 2008). At a closer look, the degree of materiality this world offers led toward urban sprawls, mega-agglomerations that are fertile grounds for apocalyptic catastrophes. While disasters have been potentiated by the introduction of unpredictable technology, it is equally true that humans are imprudently changing the ecosystem and the environment according to the mandate of profit maximization (Klein, 2014; Koch, 2011; Storm, 2009). The paradoxical situation lies in the fact that in the name of progress the instrumental manipulation not only obscures more than it clarifies but in some cases is the precondition toward disaster (Quarantelli, 2003).

To wit, Kai Erickson, an authoritative voice in disaster studies, sets forward an interesting discussion around the ideological discourses of traumatized victims in the post-disaster contexts. One of the motivations for survivors just after their homes are destroyed or their relatives killed is the hope that after all they are still alive. This means that disasters not only change the cosmology of the self but its relations with Others (Erikson, 1994).

For the public, humanitarian disasters are the product of corruption, civil wars and other types of evils, which deserve to be eradicated by the officials. By this way, the concepts of disaster and evil seem to be associated with material deprivations. Over the centuries, theologians and philosophers questioned the existence of God in view of his impossibilities to deter disasters, calamities and injustice. Even the Lisbon quake that whipped Portugal in 1755 changed the paradigm of a theocentric society toward more rational forms of thinking. This event left almost 70,000 victims and millions in material losses, questioning the conceptual doctrine of the Catholic Church. In opposite to what the Bible says, philosophers as Voltaire or Rousseau called for a modern humanism that was the background for the rise of modern seismology. Disasters are not the token of the God's plan, but the aftermaths of geological forces (Connell, 2001; Dynes, 1998, 1999). Understanding the dynamic of earth and how these forces move is the main goal of Science to make the life of people safer (Dalhammer & Tierney, 1996).

THE NARRATIVES OF DISASTERS

Disasters place the elite between the wall and the deep blue sea. In fact, disasters are certainly understood as "unexpected events" that cause material loss and deaths. The credibility of officialdom—just after the disaster takes hit—is notably undermined. The state and the officials were in charge of caring for their citizens. Hence, disasters lead toward serious epistemological and political changes, paving the ways for the rise of a new order. Doubtless, disasters not only defy the early established discourse but the logic of instrumentality from where capitalism operates. In a seminal book published in 2010, Douglas Kysar (2010) calls the attention to the limitations of Western rationality and the search for objectivity the modern science often pursues. In sharp contrast with Cass Sunstein's approaches, which are oriented voluntarily or not to an instrumental rationality, Kysar places the problem of environmentalism as a major threat for capitalist society. In this respect, as Kysar puts it, neither the cost-benefit analysis nor the precautionary principle contemplates the complex nature of the environment, as well as the controversial demarcation of uncertainness and certainness. The risk-related statistics bespeak of the certain hypothetical problem, evaluating often information scientists have carefully and selectively picked up. However, there is a dark side of the disaster, which remains inexpugnable for scientists and policy-makers. The theory of chaos teaches that randomness is a key factor of the system, placing the hegemony of reasoning into question. The risk is engendered from the future without a serious basis to what extent it will take place, in reality, some later day.

With this backdrop, Kysar holds that states play an ambiguous role in the mitigation of disasters. He puts the example of Hurricane Katrina, which is self-explanatory. Once the hurricane took landfall, the authorities who were part of Bush's administration blamed the nature for the victims and the material losses. The hurricane was responsible for the number of casualties and devastation in New Orleans. However, with the passing of days, the same authorities assumed the entire liability for the successful evacuation of survivors and the relocation of those American citizens who were trapped in the city. This exhibits the ambiguous nature of "precautionary principle", which is often manipulated by the political power, Kysar adds. While the course of actions is oriented to past, politicians blame others for their faults, but this does not happen within the borders of precautionary doctrine, which means when politicians should move resources to save lives (in a hypothesized future). This represents one of the valid reasons why—as Kysar noted—the problem of climate change remains unresolved.

Understanding Resilience: An Anthropological Viewpoint

The term "resiliency" was originally coined by Viktor Frankl, who bore the tragedy of Nazi Germany and the tough conditions of a concentration camp. The term denotes the psychological capacities of self to overcome adverse situations or conditions of extreme vulnerability. In disaster studies, the idea of resilience helped policy-makers and scientists in the elaboration of post-recovery plans. Some scholars define resilience as a framework that stimulates a cohesive self-help for the community (Drury, Cocking, & Reicher, 2009: 67). The first point of entry in this debate revolves around difference in people's behavior once the disaster hits. In emergency evacuations, survivors tend to cooperate with others looking for communal assistance and protection. In contrast to what public opinion imagines, looting and riots are uncanny unless in the plots of disaster movies and the cultural entertainment industry. As Quarantelli puts it, disasters not only should be studied as social facts but also foster the social cohesion (Quarantelli, 1988, 1997, 2005). The different reactions organized to evacuate people from the ruins, as well as the solidarity expressed with the victims, bespeak of how collective resilience can be structured according to a psychological unity that connects to a wider in-group. As Drury et al. overtly state:

> In the London bombings, survivor's behavior was characterized by adaptive features, such as order, solidarity and mutual aid rather than the dysfunctional individualism and panic that characterize psychosocial vulnerability. Importantly, it was the crowd itself that was the basis of the reisilience displayed by survivors. In this account, then the crowd is psychosocial resources: a sense of psychological unity with other during emergencies is the basis of being able to give and accept support, act together with a shared understanding of what is practically and morally necessary, and see other's plight as linked to our own rather than counterposed. (2009: 85)

A profound, coherent and integrative paper authored by Haigh and Amaratunga suggests that present research in built environment or collective resilience is obscured by ill-defined "disciplinary base". For these scholars, resilience can be comprehended as "the capacity of a system, community or society potentially exposed to hazards to adapt, by resisting or changing in order to reach and maintain an acceptable level of functioning and structure" (Haigh & Amaratunga, 2010: 14).

Nonetheless, one of the problems in the disaster's complexity is related to the fact that sometimes it is very hard to determine where a disaster begins and ends. Starting from the premise that more vulnerable classes should be protected from disasters, Haigh and Amaratunga say:

> The process of disaster management is commonly as two-phase cycle, with post-disaster recovery informing pre-disaster risk reduction and vice-versa. The disaster management cycle illustrates the ongoing process by which governments, business and civil society plan for and reduce the impacts of disasters, react during and immediately following a disaster, and take steps to recover after a disaster has occurred. (Haigh & Amaratunga, 2010: 17)

The question of poverty, as well as the humanitarian disasters, will be discussed in the next section.

Poverty in Perspective

Home is one of the sites where the person is really socialized not only as a child but also in the adulthood. Home represents an important emotional side of life-span; in Erikson's terms, a place where a person is a person. This section explores the contributions of Erikson in the fields of humanitarian disasters, as the case of Ojibwa, an Indian reservation located near Grassy Narrows in Canada. Considering the high-prevalence of murders induced by alcohol over-consumption, he argues a social disaster is gradually emerging in Grassy Narrow. Like a tornado or other natural catastrophe, social fragmentation and alcohol drinking are blurring the liaisons of families with their neighbors. In the winter of 1970, Erikson witnessed the Wabigoon River gradually being contaminated with mercury. Those natives who live in reservations were unaware that they were exposed to this poison for decades. Effects on inhabitants' health were far from being observable in the short run. Later, the clinical symptoms were associated with mental problems, clumsiness, memory loss, impaired visions and finally to depression. Per his field notes, Erikson reminds that:

> To make matters worse, the very fear that mercury generates can act to stimulate the real thing. I do not simply mean that apprehension about this or that symptom of mercury poisoning can help provoke its appearance, as is certainly true enough, but that dread itself has among its behavioral by-products the kind of disequilibrium, depression, memory loss and volatily that mercury is known for. (Erikson, 1994: 36)

This moot point raises a pungent point, was this poison responsible for the social pathologies linked to alcohol abuses Erikson saw at a first glance? Undoubtedly, the mercury wreaked havoc in the inhabitants of this little town. However, toxic poison was not the only threat this community certainly faced. White's colonization pushed hosts to change their ancestral customs and form of tribe organization. This progressive disintegration of clans in combination with difficulties to track down and hunting made day-to-day survival troublesome for Grassy Narrows's Natives.

As explained, Erikson brings in reflection to what extent our current understanding of disaster still is correct. The Spanish Flu outbreak that whipped the world between 1918 and 1919 killed a considerable part of the population in this reserve. Even though a recovery could be noticed and the population climbed from 178 in 1917 to 242 in 1949, epidemic resulted in a genuine shock for the old-shamans who were hand-tied to heal infected people. Not only was their reputation seriously jeopardized but this also represented a serious declination for religion. The white's man-God embodied in Christianity was adopted by Natives reinforcing the previous process of education accomplished by Canadian Authorities from 1870 to 1890. "The Canadian Government had been charged by a treaty signed in 1873 to provide education to the Objibwa people, and it elected to meet that obligation by building residential schools, many of them run by missionaries at some distance from the reserves themselves. Whatever their intent, the effect of these schools was to separate children from their families for long stretches of time and to help strip them of their language, their native skills, their religion, and their very identity as Indians" (Erikson, 1994: 45).

For Natives in America, lands play an important role in the configuration of their own cosmology because they converge with the spirit of ancestors. The lands bring security to dwellers because they are interrelated to the soul of rivers and mountains. Whenever persons change home, they should be sure they would be welcomed. Otherwise, bad spirits impinge the families encouraging vicious, violence, and other types of calamities. The involuntary migration—managed by the Canadian Government—toward another reserve resulted in serious pathological problems for aboriginal tribes not only in Canada but also in the US. Erikson realized that alienation and white hegemony seem to be key factors that lead these striking narrated stories of poverty, exclusion and self humiliation in a difficult position, in a real humanitarian disaster.

Similar to this above-illustrated story seems to be the case of the life of poorer Haitians in Immokalee, a little town situated southward Everglades in Florida, US. This site is well recognized as a part of the country that attracts many illegal immigrants for the harvests—farmers or chronic migrants who roam from one site to another in quest of an opportunity for social upliftment. Bereft in conditions of misery and continuous necessities, these migrants are discriminated by residents, even compared with animals or with "excrement". On March 22, 1981, *Miami Herald* referred to these newcomers as "Haitian Stampede". In the past, Haiti was one of the most prominent and important colonies of this hemisphere; now this country seems to be one of poorer ones with more than 80% of their lands uncultivated, drained or exhausted. With basis on multiple marriages, Haitian's customs notably contrast with lay Americans. Very hard to define, their religion is, of course, a mixture of French Catholicism, Vodun of Africa and evangelical Protestantism. They often live in overcrowded settlements wherein a family formed by five or more members share a small-sized household of 176 fleets and 16 foot-rooms.

To cut the long story short, Haitian peasants weekly send remittances to their relative in Haiti; many of them are sick or living in pauperism. The disaster surfaced whenever Fred's (a company aimed at sending money from US soil to the island) was declared in bankruptcy because of the embezzlement of funds of one of the owners. Erikson collected a bunch of striking chronicles of peasants who suffered the death of a relative, a father, a mother or even a son. This fraud was ultimately condemned by the courts of Florida and resulted in only 30 months of prison for Fred Edenfield. Nonetheless, consequences for Haitians—even if it did not represent too much money for other peasants—were sorrowful. Psychologically speaking, affected people started to feel depression, insecurity, pain, sorrow, disorientation. Most certainly, testimonies repeat that children were hungry and without clothes at home. An event of this caliber affected the soul of Haitian peasant deeper than thought because it involves their relatives, more of them children. The vulnerabilities of these stolen people resulted in a much broader state of insecurity and uncertainty. Erikson's contributions move in two clear-cut directions. On one hand, our emotions distort the real reasons behind the events that harm us. As the cited study cases, Erikson understands that while some risks are over-valorized others whose impacts are critical are glossed over. On the other hand, their cultural representations are replicated in the threshold of time. Like poverty and labor exploitation, which daily left thousands of workers in a miserable condition of life, the power of ideology consists in tergiversating the causes of poverty inverting the logic between cause and effect. Poverty is the main cause of disasters—disasters are the result of the capitalist exploitation over lower classes. In that way, and using a blaming-other tactic, the ruling elite embrace a cynic paternalism on the poors, which is designed to protect their hegemony. As a result of this, the probabilities of the same disaster repetitions are higher.

The Nature of Pain in the Society of Fear

The turn of the century witnessed the multiplication of different risks, dangers and disasters as never before. As privilege witnesses, the inspectorship consumed a vast range of events, which spanned from earthquakes and floods to the outbreaks of the lethal virus as Ebola, swine flu and so forth. These disrupting events created what Baudrillard dubbed as "the spectacle of disaster". In this respect, each event not only replaces an older one, but it is commoditized to be gazed and consumed by a global audience. As Baudrillard writes, capitalism blurs the present and the future as the plot of Spielberg's movie *Minority Report*. The Precogs know the future, which helped the police to eradicate the crime. What happens in the present time never truly happens in the real (Baudrillard, 2002). One of the fathers of modern anthropology, Bronislaw Malinowski (1994) proposed that societies may be studied by their taboos, instead of by their norms. In the same way, we hold that societies may be understood by their fears.

Simon, in his book, *Governing through Crime*, gives a convincing explanation why fears have been historically manipulated in the US. The check and balance forces that constituted American democracy impede a total autonomy from the Executive Branch. The symbolic war on cancer, crime and finally on terrorism gave the General Attorney a central position, enhancing the power of the Executive Branch. Beyond the idea of governance lies the needs of imposing policies otherwise would be widely rejected by the rest of the Branches. The discourse behind the War on Crime articulated a political network oriented to enhance governance but running the risk of corrupting democracy (Simon, 2007). Andrew Hoskins and Ben O'Loughlin present an all-encompassing review of the intersection of television and terror. The shock produced by 9/11 paved the ways for the rise of a long-dormant culture of fear as never before. It created an emotionally distorted response to fear, which often is amplified by the media. It is important not to lose the sight of the fact that politics and politicians in specific terms face one of the worst crises of credibility. Authors go on to write:

> We propose two concepts that give us analytical leverage to understand this crisis. The first is the modulation of terror. News modulates terror by often simultaneously amplifying and containing representations of threat. News amplifies by inflating the seriousness of threats, by connecting a single threat to others, or by representing threats in vague, indefinite terms through speculation, linguistic imprecision, or loose use of numerical, quantitative indicators of terror. (Hoskins & O'Loughlin, 2009: 14)

With this backdrop, the media fulfills an influential role in social imaginary, supplanting the needs of "being there", by the pictorial story, which can be externally fabricated. As a window, the media screen gives a story, surely of an event, viewers are impossible to grasp or at the least understand. This reflexivity performs (as a bridge) a significant role cementing the grounds of politics.

Alicia Entel, an Argentinian philosopher who does not need an introduction, published in 2007 a book entitled *La Ciudad y Sus Miedos (The City and Its Fears)*. Following similar guidelines as Erikson, she focuses on the process of resilience each society develops. Of course, these reactions and answers vary on culture and the degree of economic maturation. It is safe to say psychoanalyst offers an interesting viewpoint to the problem of trauma. For Entel, the society follows the same psychological dynamics than a person. The trauma is elaborated, repressed and forgotten, while the "frightening object" is sublimated toward a third object. Indeed, our fears are emotional elaborations to process a much deeper historical trauma. Methodologically speaking, only qualitative-related methods suffice to describe how fears operate in the society. Here two assumptions should be done. On one hand, there is a dialectics between fear and discrimination. In fact, as psychoanalysis amply showed, the phobia represents a sacred fear elaborated to avoid the fragmentation of the mind (cleavage), in this case, the excision of the society. What is more important, the fearing object should be isolated, in order for the societal order to keep the stability. In this vein, as Entel puts it, the fear allows certain mechanism of racism and discrimination aimed at undermining the integrity of the Other. In parallel, the media disseminates fear-containing news in order for certain claims would not be met. In consonance with Bauman, Entel says that through the articulation of fears, citizens accept certain economic policies (fostered by neoliberalism) otherwise would be never accepted. The current worries about local crime, following her insight, correspond with the repressed trauma created by the Juntas in the 1970s. The illegal repression and the bloody practices of violence perpetrated by the State led to the profound sentiment of frustrations, which paralyzed the political will of the citizenry. In the past, people were hosted, killed and disappeared, while today, the hostage turns symbolic. This means that thousands of citizens are pressed to live without a stable job, in miserable conditions. To put this slightly in other terms, the sentiment of insecurity results from the decline of social bondage. Once again, the poverty appears as a central figure in the discussion. Entel acknowledges that the legitimacy of states is previously determined by their efficacy in addressing and resolving citizens' demands. In Argentina, which is a country characterized by a large tradition of personalism and political instability, almost 30% live under the line of poverty. In 2001, Argentina faced one of its worst economic crises. Entel assumes that there are two types of complementary frights, one diffuse, fuzzy that terrifies collectively to the society, while the other is more concrete. Both works agree that the reciprocity among the involved sides is breached by the introduction of fear. To set an example, while the fear of losing the job is materialized as a concrete potential event, what underlies behind this is the fear to the Otherness, who would be a potential competitor at the liberal marketplace. The poverty, nowadays, hides a repressed trauma generated by the physical disappearance and the horrendous crimes committed by the Juntas. Poverty seems to be a consequence, in this case of the neoliberal policies, whereas fear ideologically operates to blame Others for the current economic conditions. In the same line, pauperism exhibits the same genocidal nature of forced disappearance in the former decades. One of the successes of neoliberalism consisted in extending the belief the nation-state is inefficient to give the basic grounds for the personal achievement. Hence, for the exegetes of neoliberalism, poverty would be the result of immigration, corruption and moral degradation. Centered on the idea that ethnic minorities are presented as scapegoats, for officialdom to conserve their legitimacy, Entel concludes the bloody legacy left by the Juntas triggered unexpected effects that persisted to date.

The Ideological Core of Poverty

Undoubtedly, one of the voices in the cultural studies on poverty is Slavoj Žižek. Though the question of ideology and poverty was discussed in a great portion of his works, we carefully selected some of them because of time and space limitations. Particularly, Žižek departs from the thesis that the success of capitalism (as a cultural and economic project) depends on its ability to convince lay citizens they are free, and their decisions made in a climate of egalitarian conditions (Žižek, 2006, 2008, 2014). We think, feel and move as we are free, when indeed we are not. So, if this would be the case, where does he place poverty?

In Žižek's account, poverty—like violence—alters the borders of responsibility as well as the casuistic of the events. The tendency to exercise violence on the alterity depends on the needs of controlling the higher levels of anxiety, uncertainness and fear that capitalism produces. As functional to the ruling elite, risks derived from the principle of shortage, which is enrooted in the modern economy and the legal figure of sovereignty, which in turn is legitimated by the biopower. The excess of rationalization can be materialized in Eichmann's trial, where Arendt showed that after all a pure evil may be banal. The instrumentalization, which is based on the means-and-end logic, explains not only the crimes organized by Nazi Germany but also modern issues as terrorism and workers precaritization. The symbolic co-action imposed by violence is accompanied by other signs as charity, sympathy and the process of victimization (Žižek, 2008). In other words, the panic and shock experience by disasters, or in this point, the landscape created by the media to understand disasters, prevent lay people to reach a correct understanding of the reasons that led to the tragic event. Likewise, the ruling elite are never questioned and interrogated by their participation in the decision-making process. He eloquently alerts that ideology works as a dream, in which case, there is a symbolic surface extracted from the reality, and a credible core which is a fake. Žižek pays attention to "a false urgency", which puts poverty as the reason of all evils, as a spectacle created to amass wealth. The analysts of marketing and the advertising industry appeal to conduct campaigns of charity in order to maximize the economic profits. The urgency, the disaster serves to create a sentiment of charity intended to keep the ideological dependence of the victim and its master. Far away from reversing the situation that preceded the disaster, capitalism replicates the conditions for reinforcing the center-periphery dependency. When Hurricane Katrina hit New Orleans, thousands of victims were hosted in stadiums. This event evinced the darkest side of American inequality and the historical discrimination suffered by Afro Americans. However, this was not echoed by the media, which focused the attention on the organized looting, larceny and outpouring violence in a post-disaster context. The ideological message replicated the WASP (white, Anglo, and Protestant) racism marking the agenda of the US. The Executive Branch intervened in the region, showing the would-be moral inability of blacks to evacuate rationally in post-disaster contexts. The violence instinctively would be the result of not only the basest instincts essential in blacks but also the impossibility of them developing a rational thought. The same examples can be applied to other cases, which range from the riots in France to Islamophobia. The responsibilities of West in colonizing the World during the nineteenth century, an event that created a great part of poverty today, are never interrogated.

The triumph of capitalism rested in hiding the reasons of this founding mythical event. The power of ideology is not manifested by what it overtly says, but by what it covers.

Conclusion

It is important not to lose sight of the fact that disasters are devastating and disrupting events which put very well the society in jeopardy. In this chapter, we discussed critically the figure of poverty as an ideological construal intended to preserve the status quo and the legitimacy of the state in post-disaster landscapes. We have confronted different arguments like Erikson, Žižek and Entel. Though from contrasting angles, they agree with the ideological essence of disaster media coverage, as well as the role played by poverty inscribing consumption into a false charity. Disasters have historically accompanied mankind from its inception, but now capitalism has enthusiastically commoditized them as real spectacles packaged to entertain a global spectatorship. This chapter delved into the state of vulnerability catastrophes often accelerate, and the blaming-other tactics of politicians to keep their privileges. It signals to assign particular cases as the valid and universal explanation of the facts. Though poverty still is one of the ethical dilemmas of Western civilization and capitalism, its ideological transformation reminds poors are responsible for disasters. The coverage of media in these types of events goes in this direction.

References

Aguirre, B. E., & Quarantelli, E. H. (2008). Phenomenology of Death Counts in Disasters: The Invisible Dead in the 9/11 WTC Attack. *International Journal of Mass Emergencies and Disasters, 26*(1), 19–39.
Baudrillard, J. (2002). La Violence du Mondial. In *Power Inferno* (pp. 63–83). Paris, Galilee. Translated to English at www.ctheory.net/text_file
Bauman, Z. (1990). Modernity and Ambivalence. *Theory, Culture & Society, 7*(2–3), 143–169.
Bauman, Z. (2006). *Liquid Fear*. Cambridge: Polity Press.
Bauman, Z. (2013). *Liquid Modernity*. New York: John Wiley & Sons.
Beck, U. (1992). *Risk Society: Towards a New Modernity* (Vol. 17). London: Sage.
Beck, U. (1996). World Risk Society as Cosmopolitan Society? Ecological Questions in a Framework of Manufactured Uncertainties. *Theory, Culture & Society, 13*(4), 1–32.

Cannon, T. (2008). Vulnerability, 'Innocent' Disasters and the Imperative of Cultural Understanding. *Disaster Prevention and Management: An International Journal*, 17(3), 350–357.

Castel, R. (2006). *La Inseguridad social: ¿Qué es estar protegido?* Buenos Aires: El Manantial.

Connell, R. (2001). *Collective Behavior in the September 11, 2001. Evacuation of The World Trade Center.* Preliminary Paper # 313. Disaster Research Center, Universidad de Delaware, Estados Unidos.

Cronin, A. K. (2015). ISIS Is Not a Terrorist Group: Why Counterterrorism Won't Stop the Latest Jihadist Threat. *Foreign Affairs*, 94, 87.

Dalhammer, J., & Tierney, K. (1996, April). *Rebounding from Disruptive Events: Business Recovery Following the Northridge Earthquake.* Annual Meeting of the North Central Sociological Association, Cincinnati, Ohio.

Drury, J., Cocking, C., & Reicher, S. (2009). The Nature of Collective Resilience: Survivor Reactions to the 2005 London Bombings. *International Journal of Mass-Emergencies and Disasters*, 27(1), 66–95.

Dynes, R. (1998). *Seismic Waves in Intellectual Currents: The Uses of the Lisbon Earthquake in 18th Century Thought.* Preliminary Paper # 272. Disaster Research Center, Universidad de Delaware, Estados Unidos.

Dynes, R. (1999). *The Dialogue Between Voltaire and Rousseau on the Lisbon Earthquake: The Emergence of Social Science View.* Preliminary Paper # 293. Disaster Research Center, Universidad de Delaware, Estados Unidos.

Dynes, R. R. (2002). *The Importance of Social Capital in Disaster Response.* UDEL Repository. Retrieved January 1, 2019, from http://dspace.udel.edu/bitstream/handle/19716/292/PP%20327.pdf?sequence=1

Dynes, R., & Quarantelli, E. L. (1975). *Community Conflict: Its Absence and Its Presence in Natural Disasters.* Columbus: Disaster Research Center.

Entel, A. (2007). *La Ciudad y sus Miedos: la pasión restauradora (The City and Its Fears: The Restorative Passion).* Buenos Aires: La Crujía Ediciones.

Erikson, K. (1994). *A New Species of Troubles: Explorations in Disasters, Trauma and Community.* New York: Norton.

Faulkner, B. (2001). Towards a Framework for Tourism Disaster Management. *Tourism Management*, 22(2), 135–147.

Ferraz, O. L. (2008). Poverty and Human Rights. *Oxford Journal of Legal Studies*, 28(3), 585–603.

Giddens, A. (1991). *Modernity and Self-Identity: Self and Society in the Late Modern Age.* Stanford, CA: Stanford University Press.

Giddens, A. (1999). Risk and Responsibility. *The Modern Law Review*, 62(1), 1–10.

Haider, M., Ahamed, N. S., & Leslie, T. (2008). Challenge for Bangladesh to Conquer Avian Influenza. *International Journal of Pharmaceutical and Healthcare Marketing*, 2(4), 273–283.

Haigh, R., & Amaratunga, D. (2010). An Integrative Review of the Built Environment Discipline's Role in the Development of Society's Resilience to Disaster. *International Journal of Disaster Resilience in the Built Environment*, 1(1), 11–24.
Hannigan, J. (1976). *Newspapers Conflict and Cooperation Content After Disaster: An Exploratory Analysis*. Preliminary Paper # 27. Disaster Research Center, Universidad de Delaware, Newark, DE.
Heidegger, M. (1962/1927). *Being and Time* (J. Macquarrie & E. Robinson, Trans.). New York: Harper.
Hoskins, A., & O'Loughlin, B. (2009). *Television and Terror: Conflicting Times and the Crisis of New Discourse*. New York: Palgrave Macmillan.
Klein, N. (2014). *This Changes Everything*. New York: Simon & Schuster.
Koch, M. (2011). *Capitalism and Climate Change: Theoretical Discussion, Historical Development and Policy Responses*. New York: Springer.
Korstanje, M. (2010). Commentaries on Our New Ways of Perceiving Disasters. *International Journal of Disaster Resilience in the Built Environment*, 1(2), 241–248.
Kysar, D. (2010). *Regulating from Nowhere, Environmental Law and the Search for Objectivity*. New Haven: Yale University Press.
Luhmann, N. (1993). *Communication and Social Order: Risk: A Sociological Theory*. New Brunswick: Transaction Publishers.
Luhmann, N. (2017). *Risk: A Sociological Theory*. Abingdon: Routledge.
Malinowski, B. (1994). The Problem of Meaning in Primitive Languages. In J. Maybin (Ed.), *Language and Literacy in Social Practice: A Reader* (pp. 1–10). Bristol: Multilingual Matters, the Open University.
Moynihan, D. P. (Ed.). (1969). *On Understanding Poverty: Perspectives from the Social Sciences* (Vol. 1). New York: Basic Books.
Pogge, T. W. (2008). *World Poverty and Human Rights*. Cambridge: Polity.
Quarantelli, E. L. (1988). Disaster Crisis Management: A Summary of Research Findings. *Journal of Management Studies*, 25(4), 373–385.
Quarantelli, E. L. (1997). Ten Criteria for Evaluating the Management of Community Disasters. *Disasters*, 21(1), 39–56.
Quarantelli, E. L. (2003). *Urban Vulnerability to Disasters in Developing Societies*. Report No. 51. Disaster Research Center, University of Delaware, Newark, DE.
Quarantelli, E. L. (Ed.). (2005). *What Is a Disaster?: A Dozen Perspectives on the Question*. Abingdon: Routledge.
Rothstein, R. (2008). Whose Problem Is Poverty? *Educational Leadership*, 65(7), 8–13.
Sartre, J. P. (1969/1958). *Being and Nothingness: An Essay on Phenomenological Ontology* (H. E. Barnes, Trans.). London: Methuen and Co. Ltd.
Simon, J. (2007). *Governing Through Crime: How the War on Crime Transformed American Democracy and Created a Culture of Fear*. Oxford: Oxford University Press.

Storm, S. (2009). Capitalism and Climate Change: Can the Invisible Hand Adjust the Natural Thermostat? *Development and Change, 40*(6), 1011–1038.
Žižek, S. (2006). *Interrogating the Real.* London: Bloomsbury Publishing.
Žižek, S. (2008). *Violence: Six Sideways Reflections.* London: Verso.
Žižek, S. (2014). *The Universal Exception.* London: Bloomsbury Publishing.

CHAPTER 8

Conclusion

The different chapters forming this book may be read separately but a shared common-thread argument remains. The society went through a new facet of capitalism where death and suffering have become as main commodities to exchange. The theory of apocalypse, adjoined to the figure of the living dead, seems to be part of this deep issue. To put the same in other terms, the idea of the undead not only coincides with the needs of audiences to consume the Other's death but the risks for humans posed by zombies as "undesired Others" (Korstanje & Skoll, 2018). Citing the TV Series *Walking Dead*, which appeals to foster a climate of total destruction, apocalypse theory speaks us of the human decay. In the middle of this mayhem, as Jameson (1991) observes, the postmodern culture rests on representations revolving around the "unconscious" orchestrating different discourses toward a deeper interrogation of the political sphere. If we toy with the belief that modernity departs from the obsession of humans to control nature, as Ulrich Beck highlights, it is interesting to question the role of technology in this ambitious plan.

Not surprisingly, Ulrich Beck (1992) writes in his book *The Risk Society* two significant axioms which merit to be discussed in this concluding chapter. On one hand, the urgency for controlling the environment opens the doors to a philosophical paradox where technology becomes itself in the generator of global risks. Chernobyl provides Beck with the example he needs. In fact, this exemplary nuclear power plant would have made the world a safer place, unless by the accident that has shocked the audiences in the globe. The same technology deployed to protect mankind creates unseen and apocalyptic risks which place humankind in jeopardy. On the other hand, and given the problem in these terms, the state of disasters does not correspond with the lack of planning but by an excess of modernity. The needs of intervention usher governments in a paradoxical situation where attempts and strategies unfolded to mitigate some risks engender new others (Beck, 1992). This point echoes the original concerns formulated by Jacques Ellul (1962, 1992) who alerted on the negative effects of technology in modern society. Technology, for Ellul, undermines not only the social trust but also affects the critical thinking leading toward a climate of exploitation and depersonalization. In this token, Andrew Feenberg (1995) warns on the role of technology (a-la-Marcuse) as the cold administrator of the capitalist order. In fact, technology ignited a new type of modern rationalization that disposes of form the worker's body at its discretion domesticating nature through the articulation of a productive system. Those supporters of robots and machines forget one important aspect of capitalism. While technology and machines operate to make our lives safer, global Western rationality overwrites other voices, knowledge and cosmologies (Feenberg, 1995). In the present book, we interrogated and reviewed the plot of some zombie novels as *World War Z*, the *Living Dead* (among others) but less attention was given to other post-apocalyptic films as *Contagion, Night of Living Dead, Resident Evil*, the *Day of the Dead* or *Noah*. The plots of these movies converge in the same direction. The humans envision in dominating the earth, expanding life and bringing new benefits to the earth, but something turns out wrong and the used technology awakes a frightening apocalypse as never before. To some extent, technology gave humans the opportunity to distinguish from the natural order but at the same time, it oppressed them—above all when ethics is derided.

Every culture developed a mythical corpus that speaks of an atemporal world where Gods and humans lived in peace or in harmony. This paradisiacal state of existence—which comes from Eastern religions—stopped once the man committed the primordial crime, sin or even confronted the Gods. Once exiled, men reproduce the conditions of salvation—one day they enjoyed—in everything and everywhere they go (Cohn, 1996; Kumar, 1996; McGinn, 1996). As Eric Ranking brilliantly put it, for some reason, the end of the world associates to the introduction of a dark technology which does not destroy the world but purges it. Moved by greed and crime, Gods have been offended by the humanity posing a major danger for the rest of creation. God disposes of a purge to correct "the original sin" while the earth is redeemed into a new sanitized facet. Rabkin reminds that the exile remembers the human impossibility to come back to the exemplary center where Gods inhabit (Rabkin, 1986). Once the purge is disposed of—in forms of fires or floods—humans are not decimated but pressed to live in a new condition. In consonance with this, Gary Wolfe coins the term "Zero Moment" to denote the return to a primitive stage of existence where humans struggled for surviving. The destruction of the world implies the eradication of old hierarchical orders and authorities, as well as the constitutional rights. While a new aristocracy arises, the lost prosperity never returns. Wolfe argues convincingly that technology plays a leading role marking the rise of civilizations and empires but when it is instrumentalized to serve the interests of the ruling elite, corruption accelerates the decline. The idea of remaking zero alludes to the crystallization of a new (grim) world where humans should adapt. Technology is the alpha and the omega of the man:

> The promise inherent in the idea of remaking zero is certainly one of the reasons this genre has survived as long as it has, and in so many guises. On the simple level of narrative action, the prospect of a depopulated world in which humanity is reduced to a more elemental struggle with nature provides a convenient arena for the sort of heroic action that is constrained in the corporate, technological world that we know. (Wolfe, 1983: 4)

The *Matrix* Saga offers an interesting example of the above-cited excerpt. Starring Keanu Reeves, Carrie-Anne Moss and Laurence Fishburne, the plot narrates a futurist world where humans serve as sources of energy for machines. Although the Matrix was originally created to serve humans, it subverts the order creating a simulation where people have a normal life. Indeed, the bodies are connected to a pod to give the Matrix the necessary electricity to function. A young programmer, Thomas Anderson, discovers—once he swallows the red pill—he was living in a dream. Morpheus says the famous phrase "welcome to the desert of the real!"

Professor William Irwin (2005) holds an interesting thesis. The Matrix reveals not only the subjectivity of reality but also the borders of pleasure which resist entering an existential sphere. Cypher betrays Morpheus to be re-conducted to the Matrix, knowing that he will live in fiction but the reality is more than he may bear. In this respect, Nixon (2005) and Korsmeyer (2005) reminds that what we see can be the illusory product of a program, but it paves the way for the rise of anti-Western rationalism that contradicts the hegemony of gazing as the best method to reach the truth. Western civilization has disposed of the gaze as a metaphor for understanding but things move from worse to worst, and the mind maddens if reality is perceived as it really is. The betrayal of Cypher explains the urgency of forgetting the past and prioritizing the pleasure of the body over the existential interrogation.

Matrix, as well as the theory of apocalypse, brings two important reflections to the foreground. First, the apocalypse should be understood as the ascendancy of reality that melts into the fictional world, which is characterized by the civilization and the immediate gratification. Humans often construct this order to tolerate the hostility of contingency. Secondly, Matrix is an illustrative example of how ideology works. Citing Lord Raglan, the life of the hero, like Neo, is uncertain for him, he is blind about his future though he should face countless dangers in his short life. Though the hero is unaware of the destiny, he accepts the challenge, testing not only his character but becoming the protector of mankind (Raglan, 2003). The theory of bottom-days and the mythical stories of Gilgamesh or Aquila speak of test the selected men should take. This is exactly a common aspect shared by the prophecy in early Israel, terrorism and Thana-capitalism. While the prophet is touched by the grace of God, the terrorist feels he or she is a selected (saint) people commanded to make the wish of the Lord. The modern audiences enjoying news related to death feel special as well renovating their loyalties to their nation-state. The same indicates in the plots of the *Night of the Living Dead* (1968) and *Day of Dead* (1985), products of the American filmmaker George Romero. The numbers of zombies outnumber humans 400,000 to 1. Dr. Sarah Bowman and Miguel Salazar escape to Fort Myers, Florida, where they join others (such as helicopter pilot John, or Bill). However, things move from worse to worst, and they are pressed to take way toward an army base located in Everglades. Once there, they are hosted by Captain Rodhes and Dr. Logan, who commands a group of scientists and militaries who surgically experiment with zombies. Survivors at the lab devote considerable time and efforts to control (if not domesticate) the wild zombies through different experiments. The tension between soldiers and scientists comes sooner than later. Captain Rodhes orders his troops to discard all experiments while Johnny—the pilot—invites Sarah for a drink. During the conversation, he confirms God is angry because of the human greed, because of the exhaustion of natural resources. The Lord, in Johnny's eyes, disposed of a purge for the earth not to be compromised while renovating mankind in a more virtuous species. The same narrative may be traced back to *The Resident Evil 2*, where Umbrella Corporation experiments with a lethal virus which spreads rapidly through the world. Alice, a perfect warrior genetically manipulated by Umbrella Co., and her friends escape from the Hive to fight against the zombies. The uses and abuses of technology brought the dead to life conforming a new devastating species. The forces of good and evil confront in a mythical battle that helps humans to expiate their sins. In the film *Contagion*, this happens when Beth Emhoff (Gwyneth Paltrow) returns to Chicago after an adventure with her former lover, and spreads a lethal virus. Her infidelity that is unknown by her husband Mitch (Matt Damon) awakens the end of the days for humanity. Finally, the health condition of Beth aggravates and she dies, and so does her six-year-old son. Mitch—like his daughter—is protected by certain immunity, while he starts an odyssey struggling against the secrecy of the Department of Homeland Security and the lack of scruples of scientists. In any case, *Contagion* should be read in opposition to *Resident Evil* in the fact that scientists are not evil-doers but real heroes who seek a cure (finding ultimately a vaccine that saves millions of lives). Both share the needs of implementing a protocol where the interpersonal contact avoids. The case of *Noah*, a film starred by Jennifer Connelly and Russell Crowe, is pretty different. The Watcher—a cast of protective angels—were punished by God when they decided to procreate with mortal women. Their offspring, the Nephilim, were giants who devoured everything in their way. God disposed of a flood to restore the order while Shemihaza (his leader) became a rock. The director Aronofsky elegantly combines some parts of Enoch's book with the Old Testament. The Watcher helps Noah to struggle against Tubal-Cain for the wish of God to be performed. God asked Noah to fabricate an ark, but he is decisively encouraged not to tell the god's plan to Cain and his offspring. Noah should select a couple of species to continue with the life. The ark contains not only the germen of life in the next world, prepared by God, but also the future of humanity in the hands of Noah. The problem is that Noah is urged to keep the secret, witnessing how the human is ruthlessly obliterated through the articulation of a universal flood. This extinction became the first genocide and Noah the architect of a new world where the sins of man are redeemed. The "shameful" Noah falls into depression and drinks all the time until God blesses the family as the new start of a re-born human race.

As stressed in the different chapters of this book, Noah's myth is the cornerstone of Thana-capitalism and, most probably, this explains why the story gained considerable traction in the recent decades. Noah's ark validates the assertion that life needs from death and vice-versa. The dark tendencies to witness the other's death are ideologically legitimated by this founding myth but it is replicated by the plots of apocalyptic movies and zombie world. While God disposes of man's extermination, Noah situates as a privileged witness of the divine plans. From that moment on, the world was divided in two, the doomed and the salved. The crucifixion of Christ not only renovated the message but in this case, humanity took revenge, not only reminding the primordial genocide in the days of Noah but killing Christ (his unique son). To cut the long story short, the narratives of apocalypse ignites an interesting discussion revolving around the role of technology dividing humans from the natural order (the creation). As Tim Ingold (2000) eloquently observed, the dwelling perspective needs from a philosophical separation of the man from the natural world. Unlike hunters and gatherers, who have fleshed out a relational perspective with nature, we are educated to feel special, as administrator of creation. The use of technology alludes to the human rationality, which, while tiding up the territory to our necessities, reactivates a sentiment of the culprit. The end-of-the-world theories connote the belief in an original sin which jeopardizes the humankind's salvation. Having said this, the book discusses to what extent modern terrorism, media and the entertainment industries are not replicating mythical narratives aimed at diminishing the understanding for the non-Western alterity. It is important not to lose the fact that terrorism as a discomposing force accelerates the decline of hospitality in the West. Last but not least, our main thesis is that the Thana-capitalist society does not intervene in the background—or the social factors—that generate social maladies but only in the resulted products. The legal abortion, in this vein, is not pretty different to capital punishment. They are not intended to rehabilitate the criminal or alerting the young mother of the best ways of preventing an unwanted pregnancy. Instead, the society liberates the forces to eradicate the undesired effects. The same applies for terrorism and the spread of Islamophobia, where the non-Western "Others" are surveilled, jailed and persecuted just in case. We live in a society, the Thana-capitalist society, which has some problems to deal with the alterity beyond the borders of digital screen. To be honest, Žižek is mistaken when he refers to 9/11 as a shocking event that brought us out of the hegemony of the virtual world. Terrorism gives the oxygen to the media to captivate our attention enmeshing us in the ideological gridlock of technology.

REFERENCES

Beck, U. (1992). *Risk Society: Towards a New Modernity* (Vol. 17). London: Sage.
Cohn, N. (1996). Upon Whom the Ends of the Ages Have Come. In M. Bull (Ed.), *Apocalypse Theory and the End of the World* (pp. 33–49). Oxford: Blackwell.
Ellul, J. (1962). The Technological Order. *Technology and Culture*, 3(4), 394–421.
Ellul, J. (1992). Technology and Democracy. In C. Mitcham (Ed.), *Democracy in a Technological Society* (pp. 35–50). Dordrecht: Springer.
Feenberg, A. (1995). Subversive Rationalization: Technology, Power and Democracy. In *Technology and the Politics of Knowledge*. Bloomington: Indiana University Press.
Ingold, T. (2000). *The Perception of the Environment: Essays on Livelihood, Dwelling and Skill*. London: Psychology Press.
Irwin, W. (Ed.). (2005). *More Matrix and Philosophy: Revolutions and Reloaded Decoded*. La Salle: Open Court Publishing.
Jameson, F. (1991). *Postmodernism, or the Cultural Logic of Late Capitalism*. Durham, NC: Duke University Press.
Korsmeyer, C. (2005). Seeing, Believing, Touching, Truth. In W. Irwin (Ed.), *The Matrix and Philosophy* (pp. 41–52). Chicago, IL: Open Courts.
Korstanje, M. E., & Skoll, G. (2018). Technology and Terror. In *Encyclopedia of Information Science and Technology* (4th ed., pp. 3637–3653). Hershey: IGI Global.
Kumar, K. (1996). Apocalypse, Millennium and Utopia Today. In M. Bull (Ed.), *Apocalypse Theory and the End of the World* (pp. 233–260). Oxford: Blackwell.
McGinn, B. (1996). The End of the World and the Beginning of Christendom. In M. Bull (Ed.), *Apocalypse Theory and the End of the World* (pp. 75–108). Oxford: Blackwell.
Nixon, D. M. (2005). The Matrix Possibility. In W. Irwin (Ed.), *The Matrix and Philosophy* (pp. 28–40). Chicago, IL: Open Courts.
Rabkin, E. (1986). Introduction: Why Destroy the World? In E. Rabkin, M. Greenberg, & J. Olander (Eds.), *The End of the World* (pp. vii–vxv). Carbondale: Southern Illinois University Press.
Raglan, F. R. S. (2003). *The Hero: A Study in Tradition, Myth and Drama*. New York: Courier Corporation.
Wolfe, G. K. (1983). The Remaking of Zero: The Beginning of the End. In E. Rabkin, M. Greenberg, & J. Olander (Eds.), *The End of the World* (pp. 1–19). Carbondale: Southern Illinois University Press.

Filmography

Contagion. (2011). Steven Soderbergh. 106 Minutes. Participant Media, US.
Night of the Living Dead. (1968). George Romero (Dir). 96 Minutes. Image Ten, US.
Noah. (2014). Dalen Aronosfky (Dir). 138 Minutes. Regency Enterprise, US.
Resident Evil 2. Apocalypse. (2004). Alexander Witt (Dir). 94 Minutes. Constantin Film, US.
The Day of the Dead. (1985). George Romero (Dir). 100 Minutes. Dead Films, US.

INDEX

NUMBERS AND SYMBOLS
9/11, xii, 1, 33, 34, 39, 46, 51, 57, 63–66, 68, 70–77, 103, 108, 115, 116, 134, 149

A
Abortion, xi, xii, 46, 53, 56–59, 148
Agamemnon, 117
Anthropology, x, 2–4, 9, 15, 19, 21–24, 38, 45, 82, 85, 87, 93, 105–106, 123, 129–131, 134
Anti-Christ, xi, 1, 15, 113
Anti-hospitality, xii, 59
Apocalypse, vii, ix–xi, 1–16, 19, 27, 39, 143, 144, 146, 148

B
Baudrillard, J., 64, 65, 71, 77, 106, 108, 134
Beck, U., 1, 38, 103, 108, 124, 125, 143, 144
Bernstein, R., 105

Bottom-days, x, 2, 4, 9, 11, 15, 19, 36, 82, 119, 146
Brooks, M., 21, 34–39

C
Campbell, J., x, 3, 15
Cannibalism, 21, 33
Capitalism, x, xiii, 21, 23, 25–27, 32, 47–49, 51, 52, 56, 65, 66, 71, 75, 77, 81–84, 86, 94, 95, 103, 106–110, 118, 119, 124–126, 128, 134, 136–138, 143, 144
China, xi, 34, 36, 37
Cinema, 2, 13–15, 19, 21–24, 38, 40, 94, 109
Cold War, 46, 68
Colonialism, 21, 32, 45, 46, 82, 83, 88–91, 94, 95
Concentration camps, 83, 107, 111, 129
Constitutional liberties, 11
Contagion, 144, 147

Cuba, xi, 25, 35, 36
Culprit, 148
Culture, vii, ix–xiii, 3, 4, 6–8, 10–12, 15, 19, 20, 22–24, 31–33, 36, 39, 45, 48, 50, 51, 56, 57, 59, 68, 75, 84, 87, 88, 97, 103, 105, 109, 111, 113, 118, 119, 125, 127, 134, 135, 143, 144

D

Dark tourism, 82, 83, 107, 109, 110
Death, x–xii, 4, 5, 8, 20, 21, 32–34, 38, 39, 52, 53, 56, 57, 77, 83, 95–97, 104, 106–111, 117, 118, 128, 133, 143, 146, 148
Democracy, ix–xi, 2, 22, 26, 27, 35, 36, 46–48, 57, 67, 71, 72, 75, 77, 82–84, 88, 104, 134
Destruction, xi, 8, 11, 14, 20, 72, 104, 110, 143, 145
Devastation, 129
Development, 81, 82, 84–92, 94, 95, 108, 124
Disasters, ix, x, xiii, 12, 13, 70, 83, 95, 97, 107–112, 123–138, 144
Dominant hegemony, ix
Domination, x, 53, 55, 70
Douglas, M., x, 3, 4, 15, 30

E

Eden, ix, xiii, 15, 40, 111
Eliade, M., x, 2, 5–8, 15
Ethics, 24, 81, 82, 86, 91–92, 95, 96, 104, 111, 124, 144
Ethnocentrism, 22, 45, 46, 48, 49, 85, 115
Ethnology, 2, 3, 15
Europe, 9, 22, 40, 45, 58, 64, 66, 88, 89, 95, 103, 106, 113, 118, 119
European nations, 45, 53, 114

Evilness, x, 2, 97, 118
Evolution, xi, 9, 40, 45, 53, 68, 82
Exploitation, 25, 32, 33, 45, 50, 52, 53, 74, 75, 84, 87, 88, 91, 94, 96, 115, 133, 144

F

Fear, vii–xii, 1, 10, 11, 15, 19, 20, 29–33, 38, 39, 46, 48, 51, 56–59, 64–67, 70, 71, 74, 76, 77, 103, 117, 118, 123–138
Flood, ix, 110, 134, 147

G

Generation, 7, 10, 39, 40, 54
Genocide, 50, 84, 103–119, 123, 147, 148
Giddens, A., 38, 55, 56, 103, 126

H

Heroism, 4, 5, 30
History, 4–7, 9, 23, 24, 26–29, 32, 34, 37, 39, 67, 84, 86, 92, 104, 105, 111–113, 115, 124
Holocaust, xii, 103, 105–106, 111, 114, 115
Hospitality, xi–xiii, 40, 46, 52, 56–58, 71, 91, 117, 148
Hunger Games, 38, 95, 109, 110

I

Ideology, 22, 24–27, 39, 49, 55, 74–77, 87, 90, 104, 111, 124, 133, 136–138, 146
Ingold, T., 148
Instrumentalization, xii, 76, 81, 104, 106, 137, 145
Israel State, 103–119

INDEX

K
Kinship, 4
Knowledge, viii, 2, 3, 6, 55, 77, 93, 125, 126, 144
Korstanje, M., vii, x, 10, 15, 19, 34, 38, 40, 47, 48, 50, 57, 58, 63, 67, 69, 71, 77, 82, 83, 85, 90, 91, 94, 95, 103, 106, 115, 117, 123, 143

L
Lévi-Strauss, C., 3
Literature, 3, 15, 21–24, 28–30, 53, 87, 116, 124, 125
Living dead, 19, 21, 30–36, 38, 39, 143
Luhmann, N., 103, 126, 127

M
MacCannell, D., 86, 87, 96
Machines, viii, 77, 144, 145
Mankind, vii–ix, xiii, 2, 5, 8, 11, 12, 14, 15, 20, 21, 37, 39, 71, 111, 138, 144, 146, 147
Matrix, vii, viii, 3, 22, 23, 53, 86, 145, 146
Millennium, vii, 12
Mobilities, 72, 77, 85–87, 90, 95
Morbid consumption, x, 15, 81–97, 103
Myth-builder, x, 15
Myths, x, xiii, 2–11, 15, 19, 27, 28, 33, 110, 111, 116, 117, 148

N
Nazi Germany, 103, 115, 118, 129, 137
Noah, xiii, 110, 111, 144, 147, 148

O
Otherness, x, 85, 87, 136

P
Palestine, 36, 106, 113, 114
Paradise, xiii, 11, 36
Pleasure, ix, xi, 10, 23, 27, 38, 57, 81, 82, 96, 145, 146
Poverty, xiii, 26, 32, 52, 65, 81, 82, 84, 87, 89, 90, 92–96, 124, 125, 127, 131–133, 135–138
Predestination, 47, 49, 117–119, 126

R
Rabkin, Y., vii, 11, 113, 114, 118, 145
Racism, 51, 115, 135, 137
Rationalization, ix, 137, 144
Resident Evil, 144, 147
Risk, viii, xi, 1, 6, 29, 37, 45, 51, 57, 58, 65, 66, 69, 71, 73, 95, 103, 108, 115, 117, 123–127, 129, 130, 133, 134, 137, 143, 144
Risk society, xii, 1, 96, 103, 108
Romero, G., 19, 32, 146
Russia, 35, 114

S
Safety, 67
Salvation, 47, 50, 110, 119, 144, 148
Scatology, x, 8, 10, 15
Security, ix, xii, 29, 35, 47, 51, 52, 58, 59, 66, 67, 72, 74, 76, 90, 108, 123, 125–127, 132
Sin, xiii, 12, 15, 48, 49, 110, 114, 144, 145, 147, 148
Skywalker, 117
Slumming, 94, 95
Slum tourism, 93–95, 107
Social Darwinism, xii, 39, 49, 50, 94, 95, 97, 104, 108, 118, 119

Spectacle, 23, 65, 82, 95, 97, 106, 108–111, 137, 138
Star Wars, 116
Stereotypes, 2, 22, 27, 54, 69
Sublimation, 11, 14, 21, 40, 49, 135
Surveillance, 23, 51

T
Technology, vii–x, 11–15, 20, 22, 28, 32, 36, 39, 51, 65, 73, 87, 91, 106, 110, 127, 143–145, 147–149
Terrorism, vii, ix–xii, 2, 33, 40, 46–48, 50, 51, 56–59, 63–66, 68–77, 95, 103, 104, 108, 111, 123, 126, 134, 137, 146, 148, 149
Thana-capitalism, xii, 15, 21, 38, 39, 56, 57, 77, 95–97, 103–119, 146, 148
Totalitarianism, ix, 10, 30
Tourist gaze, 23, 82, 85, 86
Trump, D., 51, 57
Truth, viii, ix, 5, 146
Tzanelli, R., xiii, 94, 107

U
Undead, 19, 20, 35, 39, 143

Undesired Other, 45–59, 143
United States (US), ix, xi, 1, 2, 10, 14, 20, 28–30, 33, 35, 37, 46–48, 50, 51, 54, 57, 63, 64, 66–70, 72–77, 84, 85, 89, 95, 103, 104, 108, 132–134, 137
Urry, J., 23, 85, 86

V
Virus, xi, 20, 21, 33–37, 39, 95, 127, 134, 147

W
Walking Dead, 19, 39, 143
War on Terror, xii, 1, 46, 51, 63–77
Watchers, the, 147
World War Z, xi, 21, 34–39, 144

Z
Zero-patient, 34, 39
Žižek, S., vii, 26, 27, 56, 65, 92, 93, 106, 113, 115, 124, 136–138, 149
Zombies, xi, 13, 19–40, 46, 57, 143, 144, 146–148

The manufacturer's authorised representative in the EU is Springer Nature Customer Service Centre GmbH, Europaplatz 3, 69115 Heidelberg, Germany. If you have any concerns regarding our products, please contact ProductSafety@springernature.com

Printed and bound by CPI Group (UK) Ltd, Croydon, CR0 4YY

23/03/2026

02076402-0010